Numerical Analysis for Science, Engineering and Technology

Authored By

Prof. Said Gamil Ahmed Sayed Ahmed

Professor of Mathematics and Numerical Analysis, Faculty of Engineering, Zagazig University, Egypt

General:

1. Any dispute or claim arising out of or in connection with this License Agreement or the Work (including non-contractual disputes or claims) will be governed by and construed in accordance with the laws of the U.A.E. as applied in the Emirate of Dubai. Each party agrees that the courts of the Emirate of Dubai shall have exclusive jurisdiction to settle any dispute or claim arising out of or in connection with this License Agreement or the Work (including non-contractual disputes or claims).

2. Your rights under this License Agreement will automatically terminate without notice and without the need for a court order if at any point you breach any terms of this License Agreement. In no event will any delay or failure by Bentham Science Publishers in enforcing your compliance with this License Agreement constitute a waiver of any of its rights.

3. You acknowledge that you have read this License Agreement, and agree to be bound by its terms and conditions. To the extent that any other terms and conditions presented on any website of Bentham Science Publishers conflict with, or are inconsistent with, the terms and conditions set out in this License Agreement, you acknowledge that the terms and conditions set out in this License Agreement shall prevail.

Bentham Science Publishers Ltd.
Executive Suite Y - 2
PO Box 7917, Saif Zone
Sharjah, U.A.E.
Email: subscriptions@benthamscience.org

CONTENTS

Dedication

To

My Mother ....

My Father ....

My Sons and daughters ....

Said Gamil Ahmed Sayed Ahmed

PREFACE

To Mathematician around the world, all of us know well that mathematics is the mother of all other sciences and all of us know well that no progress without progress of the mother of all sciences. Also all of us know that all of us will die and the work will still alive. Our main message herein, is to respect the mathematics, especially in all Arab countries and asking principles to give attention to mathematics and whom working with it. We hope receiving our message to all around the world.

CONFLICT OF INTEREST

The authors confirm that they have no conflict of interest to declare for this publication.

ACKNOWLEDGEMENTS

Declared none.

Said Gamil Ahmed Sayed Ahmed
Professor of Mathematics and Numerical Analysis,
Faculty of Engineering, Zagazig University,
Egypt

CHAPTER 1

Errors: Theory and Application

Abstract: In this chapter, we will focus on the meaning of error and try to define the errors. Different and major types of errors, sources of errors, and efforts to reduce the errors will be introduced. Also, we will show how a small error in the beginning of computation will increase at the end.

Keywords: Absolute error, Error definition, Error sources, Functions of several variables, Random errors, Relative error, Systematic errors, Types of error.

1. INTRODUCTION

Let us start the present chapter by the fact that nothing in our life is hundred percent complete. Another fact is that, we can say that it is never possible to measure any thing exactly. It is acceptable to make the error as small as possible.

Error exists in everything in our daily life, but how this error affects the results and decision makers? This is the question that should be answered clearly.

The present chapter will deal mainly with the errors in calculus, integration, numerical methods for solving linear system of equations and the error when solving nonlinear equation using any iterative methods and/or system of nonlinear systems of equations. Also, it deals with the error associated with the approximation theory.

2. THE IDEA OF ERROR

Let us at first mention three different words, *exact or true*, *measured*, and *accepted value*. The first word means the actual or real value without any error, the second means measuring using hand made tool therefore, may be it will contain some difference between the measured value and the exact one, and the third one means how can we accept data as it is?

It is a wrong statement that one can define the difference between the measurement and the accepted value by the error.

Such accepted values are just measurements made by other people which have errors associated with them as well, from the accepted values, the error starts appearing.

3. CLASSIFICATION OF ERROR

As defined previously that the error comes from the measured and the accepted values, generally, one can classify error into two types *systematic* and *random* [1-3].

3.1. Systematic Errors

A systematic error is the error that *can be controlled* by any means.

3.2. Random Errors

In this type of error *no one can control* the existence or propagation of the error.

3.3. Absolute Error

The absolute error is the absolute difference between exact and approximate values, defined as:

$$\text{Abs}(x) = \left| x_{\text{exact}} - x_{\text{approximated}} \right|$$

3.4. Relative Error

The relative error is the absolute difference between exact and approximate values divided by the exact value, defined as:

$$\text{Rel}(x) = \left| \frac{x_{\text{exact}} - x_{\text{approximated}}}{x_{\text{exact}}} \right|$$

Example

Assume that the exact value of the variable x equals 0.95 and the computed or approximated values were 0.9483 correct to four decimal places, then:

$$\text{Abs}(x) = \left| x_{\text{exact}} - x_{\text{approximated}} \right|$$
$$= \left| 0.95 - 0.9483 \right|$$
$$= 0.0017$$

And

$$\mathrm{Rel}(x) = \left| \frac{x_{\text{exact}} - x_{\text{approximated}}}{x_{\text{exact}}} \right|$$

$$= \left| \frac{0.95 - 0.9483}{0.95} \right|$$

$$= 0.0017894737$$

4. ERROR SOURCES

The sources of existence of errors come from different ways, in what fellow we will try to introduce some of these sources.

4.1. Error from Input Data

The input data comes from direct measurements in the laboratory or in the field and hundred percent there will be some error in these measurements.

4.2. Device Error

The error in devices which are used in the laboratory or in a field designated with some tolerance, is called device error.

4.3. Algorithmic Error

In mathematical physics, there are so many empirical relations which are usually used and derived mainly on the basis of experiments. Using these relations in computations, will lead some errors, which is called algorithmic usage error.

4.4. Truncation Error

The word truncation has a wide range of physical and mathematical meaning [4, 5].

4.4.1. Mathematical Meaning of Truncation

Let us start with its mathematical meaning, and assume that we have a series of expansions of the exponential function. lets evaluate $\exp(1.5)$ once up to three terms and once more up to four terms, and let us see if there will be a major difference or not?

$$\exp(x) = 1 + x + \frac{x^2}{2!} + \frac{x^3}{3!} + + \frac{x^4}{4!} + + \frac{x^5}{5!} + \ldots\ldots$$

For three terms:

$$\exp(x) = 1 + x + \frac{x^2}{2!} = 1 + (1.5) + \frac{(1.5)^2}{2!}$$

$$= 3.62500$$

For four terms:

As we can see, there is a difference. Lets ask ourself another question, what will be the expected error between the exact value and the evaluated results above?

The exact value is $\exp(1.5) = 4.48169$. The absolute error between exact and the three terms approximation is given as:

$$\text{Abs}(x) = \left| \exp(1.5)_{\text{exact}} - \exp(1.5)_{\text{approximated 3 terms}} \right|$$

$$= \left| 4.48169 - 3.62500 \right|$$

$$= 0.85669$$

The absolute error between exact and the four terms approximation is given as:

$$\text{Abs}(x) = \left| \exp(1.5)_{\text{exact}} - \exp(1.5)_{\text{approximated 4 terms}} \right|$$

$$= \left| 4.48169 - 4.18750 \right|$$

$$= 0.29419$$

From all the above computations, one can conclude that taking a large and sufficient number of terms in the expansion will lead to nearly exact value but not exact at all.

4.4.2. Physical Meaning of Truncation

To understand the real meaning of physical truncation, let us focus on the physical meaning of truncated error by taking the following practical example:

In the field of petroleum, when one wants to make a simulation of the oil pipe in the earth, the pipe is too long and deep and if one wants to consider the domain of interest, actually, the domain will be from the surface of the ground up to negative infinity, but do you think is it acceptable in modeling and simulation?

The answer is no but they take long enough pipe depth and what is called in the field of modeling and simulation, the truncation technique.

This truncation will definitely lead to some error in the computated results but researchers overcome this error by different techniques.

4.5. Computation Error

This type of error comes from computation and to understand it well let us take the following example. Suppose that we have two exact values x_{exact} and y_{exact}, and the approximate values for the same variables are $x_{approximate}$ and $y_{approximate}$.

Also assume that the errors are as follows:

$$E_x = x_{exact} - x_{approximate}$$

$$\Rightarrow$$

$$x_{exact} = x_{approximate} + E_x$$

$$E_y = y_{exact} - y_{approximate}$$

$$\Rightarrow$$

$$y_{exact} = y_{approximate} + E_y$$

Adding or subtracting E_x and E_y, leads to:

$$\left| E_x \pm E_y \right| \leq \left| E_x \right| \pm \left| E_y \right| \tag{1}$$

But if one adds or subtracts the two exact values, then let's have a look at the result and its relation to the approximate values and the errors:

$$x_{exact} \pm y_{exact} = \left(x_{approximate} + E_x\right) + \left(y_{approximate} + E_y\right)$$

$$x_{exact} \pm y_{exact} = \left(E_x + E_y\right) + \left(x_{approximate} + y_{approximate}\right) \tag{2}$$

Note that the first term in equation (2) is the right hand side of equation (1), and this means that the error will be accumulated due to addition or subtraction. The same comment is valid during other operations, like multiplication and division.

5. ABSOLUTE ERROR FOR FUNCTION OF SEVERAL VARIABLES

Assume that we have a function of several variables [6-13], as follows:

$$f = f\left(x_1, x_2, x_3, \ldots\ldots\ldots, x_n\right) \tag{1}$$

Assume that the error in each variable, in respective manner as $\Delta x_1, \Delta x_2, \Delta x_3, \ldots, \Delta x_n$, then the error that corresponds to these errors, will take the following form:

$$\Delta f = \frac{\partial f}{\partial x_1}\Delta x_1 + \frac{\partial f}{\partial x_2}\Delta x_2 + \frac{\partial f}{\partial x_3}\Delta x_3 + \ldots\ldots + \frac{\partial f}{\partial x_n}\Delta x_n \tag{2}$$

Taking the absolute for both sides of equation (2), and taking the concept of inequalities into consideration, we have:

$$|\Delta f| = \left|\frac{\partial f}{\partial x_1}\Delta x_1 + \frac{\partial f}{\partial x_2}\Delta x_2 + \frac{\partial f}{\partial x_3}\Delta x_3 + \ldots\ldots + \frac{\partial f}{\partial x_n}\Delta x_n\right| \tag{3}$$

Then

$$|\Delta f| \le \left|\frac{\partial f}{\partial x_1}\Delta x_1\right| + \left|\frac{\partial f}{\partial x_2}\Delta x_2\right| + \left|\frac{\partial f}{\partial x_3}\Delta x_3\right| + \ldots.. + \left|\frac{\partial f}{\partial x_n}\Delta x_n\right|$$

$$\tag{4}$$

$$= \left|\frac{\partial f}{\partial x_1}\right||\Delta x_1| + \left|\frac{\partial f}{\partial x_2}\right||\Delta x_2| + \left|\frac{\partial f}{\partial x_3}\right||\Delta x_3| + \ldots + \left|\frac{\partial f}{\partial x_n}\right||\Delta x_n|$$

Hence, the maximum error occurred for the function that will correspond to the errors in the contained variables, will take the following form:

$$\left|\Delta f\right|_{max} = \left|\frac{\partial f}{\partial x_1}\right|\left|\Delta x_1\right| + \left|\frac{\partial f}{\partial x_2}\right|\left|\Delta x_2\right| + \left|\frac{\partial f}{\partial x_3}\right|\left|\Delta x_3\right| + ... + \left|\frac{\partial f}{\partial x_n}\right|\left|\Delta x_n\right| \qquad (5)$$

And the maximum relative error will be:

$$\text{Max. R. E.} = \frac{\left|\Delta f\right|_{max}}{\left|f\right|} = \left|\frac{\partial f}{\partial x_1}\right|\left|\Delta x_1\right| + \left|\frac{\partial f}{\partial x_2}\right|\left|\Delta x_2\right|$$

$$\qquad (6)$$

$$+ \left|\frac{\partial f}{\partial x_3}\right|\left|\Delta x_3\right| + ... + \left|\frac{\partial f}{\partial x_n}\right|\left|\Delta x_n\right|$$

Example

Find the maximum relative error for the function $f(x, y, z) = \frac{xy}{z}$, given that

$x = 1 \pm 0.01$

$y = 2 \pm 0.03$

$z = 3 \pm 0.04$

Solution

From the given values for the variables, one can conclude the following numeric values:

$x = 1 \pm 0.01 \Rightarrow x = 1 \quad \& \quad \Delta x = \pm 0.01$

$y = 2 \pm 0.03 \Rightarrow y = 2 \quad \& \quad \Delta x = \pm 0.03$

$z = 3 \pm 0.04 \Rightarrow z = 2 \quad \& \quad \Delta x = \pm 0.04$

The next step is to write the general formula for the maximum relative error:

$$\text{Max. R. E.} = \frac{|\Delta f|_{\text{max}}}{|f|} = \left|\frac{\partial f}{\partial x}\right||\Delta x| + \left|\frac{\partial f}{\partial y}\right||\Delta y| + \left|\frac{\partial f}{\partial z}\right||\Delta z|$$

We see that, we have to prepare ourselves before computations, *i.e.*, we need to find all the partial derivatives that appear in the maximum relative error formula:

$$f_x(x, y, z) = \frac{y}{z}$$

$$f_y(x, y, z) = \frac{x}{z}$$

$$f_z(x, y, z) = -\frac{xy}{z^2}$$

The next step is to evaluate the function and the obtained partial derivatives at the point $(x, y, z) = (1,2,3)$, therefore:

$$f(x, y, z)|_{(1,2,3)} = \frac{2}{3}$$

$$f_x(x, y, z)|_{(1,2,3)} = \frac{2}{3}$$

$$f_y(x, y, z)|_{(1,2,3)} = \frac{1}{3}$$

$$f_z(x, y, z)|_{(1,2,3)} = -\frac{2}{9}$$

$$\text{Max. R. E.} = \frac{|\Delta f|_{\text{max}}}{|f|} = \left|\frac{\partial f}{\partial x}\right||\Delta x| + \left|\frac{\partial f}{\partial y}\right||\Delta y| + \left|\frac{\partial f}{\partial z}\right||\Delta z|$$

$$= \left(\frac{2}{3}\right)(0.01) + \left(\frac{1}{3}\right)(0.03) + \left(\frac{2}{9}\right)(0.04)$$

$$= \frac{41}{600}$$

SUPPLEMENTARY PROBLEMS

1-Find the value of $\exp(x)$, for $x = 1.2$, using expansion of the exponential function up to the following cases:

1-1 Up to 4-terms

1-2 Up to 6-terms

1-3 Up to 8-terms

Then, compare the results with the exact value, and determine:

1-4 Absolute error in each case

1-5 Relative error in each case

2-If Taylor's series for

$$\cos x = 1 - \frac{1}{2!}x^2 + \frac{1}{4!}x^4 - \frac{1}{6!}x^6 + - \ldots + \frac{(-1)^n}{(2n)!}x^{2n} + R_n(x)$$

Evaluate the function at $x = \dfrac{\pi}{3}$, $n = 0$ to 6

3-Suppose we have forward difference table of the following form:

x	$f(x)$	$\Delta f(x)$	$\Delta^2 f(x)$	$\Delta^3 f(x)$	$\Delta^4 f(x)$
0.0	0.000				
0.2	0.203	0.203			
0.4	0.423	0.220	0.017		
0.6	0.684	0.261	0.041	0.024	
0.8	1.030	0.346	0.025	0.044	0.020
1.0	1.557	0.527	0.181	0.096	0.052
1.2	2.572	1.015	0.488	0.307	0.211

4-Suppose that the errors in the measured data $f(x)$ are as follows:

x	0.0	0.2	0.4	0.6	0.8	1.0	1.2
$E(f(x))$	0.001	0.012	0.002	0.006	-0.007	0.018	0.009

Re-construct the forward difference table for the errors to show the error propagation in the forward difference table.

5-Assume that we have:

$$x_{exact} = 3.0, y_{exact} = 3.6 \text{ and } x_{approx} = 2.9990 \& y_{approx} = 3.5998$$

Check whether $|E_x \pm E_y| < |E_x| \pm |E_y|$ is verified or not?

6-Find the round-off error between the two successive results in the following case:
$$x_{n+1} = 3.10685 \quad x_n = 3.08254$$

7-If we have: $x_{exact} = 1.6, y_{exact} = 1.9$ and $x_{approximzte} = 1.6001 \& y_{approximzte} = 1.9001$
Check whether $|E_x \pm E_y| < |E_x| \pm |E_y|$ is verified or not?

8-Find the round-off error between the two successive results in the following case:
$$x_{n+1} = 1.553427 \qquad x_n = 1.541271$$

9-Find the round-off error between the two successive results in the following case:
$$x_{n+1} = 1.553427 \& x_n = 1.541271 \text{ correct to 4-decimal places.}$$

CHAPTER 2

Theory of Approximations

Abstract: Function approximations, interpolation, numerical differentiation and numerical integrations are very important topics in the numerical analysis. These topics are widely used and found in science and engineering applications. In the present chapter, we will present the main methods and techniques that had been developed for such topics. Differentiation and integration related together by so many relations as we will see later on. Finally, one can say that these topics are very important in our daily life, and therefore everyone should be aware of them.

Keywords: Derivative approximations, Difference tables, Interpolation, Lagrange method, Numerical differentiation, Numerical integration, Series tests, Taylor expansion.

1. INTRODUCTION

Function approximations, interpolation, numerical differentiation and numerical integrations are very important topics in numerical analysis. These topics are widely used and found in science and engineering applications.

Numerical differentiation is a very important topic in numerical analysis and found in so many applications [14]. Numerical differentiation is in relation with different topics in mathematics in general and numerical analysis in particular.

Many practical and engineering applications require numerical estimates of derivatives of functions appeared throughout dealing with such applications.

When the analytical solutions are not possible or difficult to obtain, then researchers prefer the numerical approach to find approximate values for both differentiation and integration.

Therefore, it is important to have good approximate or numerical methods to compute and manipulate derivatives and integrals.

In the next sections, we are going to present a number of methods for doing numerical integration and differentiation. Also, we are going to present a general strategy for deriving such methods.

2. DERIVATIVES FROM TAYLOR'S SERIES

2.1. First Derivative

It is well known that derivation of the derivative of any order based Taylor's series requires brief knowledge of Taylor's series. We know that Taylor requires some conditions to be satisfied with the function to be expanded, that is at least $(n+1)$ derivatives should exist to ensure finding Taylor's series.

Let us start by re-writing the Taylor's series up to the second derivatives, about the point of expansion x_0, as follows:

$$f(x_0 + h) = f(x_0) + hf'(x_0) + \frac{h^2}{2!}f''(x_0) + \dots\dots \tag{1}$$

Let us again have a look at equation (1) and neglect the terms of higher order derivatives more than the first order, we can get:

$$f(x_0 + h) = f(x_0) + hf'(x_0) \tag{2}$$

Now, let us re-write equation (2) in the following simplified form:

$$f'(x_0) \cong \frac{f(x_0 + h) - f(x_0)}{h} \tag{3}$$

By deep look to equation (3), we see that it represents the approximation of the first derivative to the given function.

Example

Use the data in Table (**1**) to estimate the first derivative of y at $x = 1.7$, use $h = 0.2$.

Hint:

The numerical values given below in Table (**1**) are for the function $f(x) = e^x$

Table 1. Numerical data.

x	1.3	1.5	1.7	1.9	2.1	2.3	2.5
y	3.669	4.482	5.474	6.686	8.166	9.974	12.182

From calculus, we know well that the function has different form, and one of these forms is the tabulated function. The numerical values in the given table are originally for the function $f(x) = e^x$.

It is important herein to remember that when it is required to find the first derivative for a tabulated function, we should choose a point in the table to be very close to the point at which we will evaluate the numerical first derivative.

Therefore, it is required to evaluate the first derivative at the point $x_0 = 1.7$, and then the best point to be chosen will be $x_0 = 1.5$. Now, let us start the solution by re-writing the formula for evaluating the first derivative based on the Taylor's series as follows:

$$f'(x_0) \cong \frac{f(x_0 + h) - f(x_0)}{h}$$

$$f'(x_0) \cong \frac{5.474 - 4.482}{0.2}$$
$$= 4.960$$

Hint

Anyone who works in the field of numerical analysis must understand the fact that he cannot expect exact results; however he must do his best to ensure best results with acceptable advanced accuracy, usually called prescribed accuracy.

2.2. Second and Higher Derivatives

The main idea of deriving a formula for the second and higher order derivatives based Taylor's series comes from re-writing two different Taylor series with two different signs.

Adding or subtracting them according to the required derivative will lead to the required formula.

Furthermore, we will start deriving the basis formula for the second derivative, and so we will start doing that by writing two Taylor series as follows [15]:

$$f(x_0 + h) = f(x_0) + hf'(x_0) + \frac{h^2}{2!}f''(x_0) + \frac{h^3}{3!}f''(x_0) + +..... \tag{4}$$

$$f(x_0 - h) = f(x_0) - hf'(x_0) + \frac{h^2}{2!}f''(x_0) - \frac{h^3}{3!}f''(x_0) + -... \tag{5}$$

By adding equations (4) and (5), leads to:

$$f(x_0 + h) + f(x_0 - h) = 2f(x_0) + h^2 f''(x_0) \tag{6}$$

Re-write equation (6) as follows:

$$f''(x_0) = \frac{f(x_0 + h) - 2f(x_0) + f(x_0 - h)}{h^2} \tag{7}$$

Equation (7) is the required formula for evaluating the second derivative of any function based Taylor's series.

Example

Use the data in Table (2) to estimate the second derivative of y at $x = 1.7$. Use $h = 0.2$

Table 2. Numerical data.

x	1.3	1.5	1.7	1.9	2.1	2.3	2.5
y	3.669	4.482	5.474	6.686	8.166	9.974	12.182

Here, it is important to remember that we should choose a point in the table to be very close to the point at which we will evaluate the numerical first derivative.

Therefore, it is required to evaluate the first derivative at the point $x_0 = 1.7$, and then the best point to be chosen will be $x_0 = 1.5$.

$$f''(x_0) = \frac{f(x_0 + h) - 2f(x_0) + f(x_0 - h)}{h^2}$$

$$f''(x_0) = \frac{3.669 - 2(4.482) + 5.474}{(0.2)^2}$$

$$= 4.475$$

It is important to remind the reader the fact that numerical differentiation formulae are not accurate well due to the truncation of the expansion terms according to the order of the derivative required.

One of the most interesting methods for numerical differentiation, is that derived from interpolation, forward, backward or central and it is found from the experience that results of numerical differentiation based interpolation are more accurate and stable than those based on Taylor's series.

Furthermore, we will start derivation differentiation formula based forward interpolation for raw data.

3. THEORY OF APPROXIMATION

One of the most interesting topics in the field of numerical analysis is the approximation theory. In most applications in nature, the original functions representing set of specified data do exist and it is so difficult to obtain and we have only some tabulated data representing it [16].

Therefore, it will be important to replace the original function by a polynomial of a prescribed order, and that polynomial will have the same properties as can be possible of the original one.

The topic of theory of approximation will be studied in some details in the next sections, before continuing the subject of the numerical differentiation.

In the next section, we will start with Taylor's polynomial and we will show the main difference between Taylor's series and Taylor's polynomial [17].

4. TAYLOR POLYNOMIAL

Suppose that we have a function defined over a certain interval. We look for a polynomial that can be used instead of the given function.

The required polynomial will have the same properties of the given function in such a way that the error between the function and the corresponding polynomial is very small and a prescribed error tolerance is allowed.

This approximation with the very small error allowance will be at the point of expansion and may increase when moving away that point.

It will be interesting if we can get a polynomial very close to the function over the interval of definition, but this topic is still point of research till that moment [18].

5. GENERAL STRATEGY OF SOLUTION

1-Assume a polynomial of a sufficient order in advance.

2-Select in advance point to be the point of expansion.

3-It is axiom that the values and derivatives of both function and the corresponding polynomial are equal at the point of expansion.

4-Equate the function values and its derivative with the values and derivatives of the assumed polynomial, leads to the unknown coefficients.

Example

Find a polynomial of degree three or less to approximate the function given by $y = \sin x$ near $x_0 = 0$, and then use this polynomial to find $\sin (0.1)$.

Solution

First of all, when solving any numerical problem, it is recommended to concentrate on what is required. The requirement in the following example is to find a polynomial of order three or less.

This means that the maximum order will not exceed three. Therefore, the first step is to assume a polynomial of order three.

$$p(x) = a_0 + a_1 x + a_2 x^2 + a_3 x^3$$

The next step is to find the derivatives up to the third order for both the function and the assumed polynomial, this will be as follows:

The polynomial

$$p'(x) = a_1 + 2a_2x + 3a_3x^2$$

$$p''(x) = 2a_2 + 6a_3x$$

$$p'''(x) = 6a_3$$

The given function

$$f'(x) = \cos x$$

$$f''(x) = -\sin x$$

$$f'''(x) = -\cos x$$

Next is to evaluate both the function and its derivatives and the polynomial and its derivatives at the point of expansion, as given in Table (3).

Table 3. The function and the approximated polynomial with their derivatives.

Function	Polynomial
$f(x) = \sin(0) = 0$	$p(x=0) = a_0$
$f'(x) = \cos(0) = 1$	$p'(x=0) = a_1$
$f''(x) = -\sin(0) = 0$	$p''(x=0) = 2a_2$
$f'''(x) = -\cos(0) = -1$	$p'''(x=0) = 6a_3$

Finally, equate each corresponding cell in each row in the above Table (3) this will lead to the values of the unknown coefficients, as follows:

$$a_0 = 0$$

$$a_1 = 1$$

$$a_2 = 0$$

$$a_3 = -1/6$$

Then, the required polynomial will take the following form:

$$p(x) = x - \frac{1}{6}x^3$$

6. LAGRANGE POLYNOMIALS

In some cases, as we mentioned before, we encounter some types of functions, called tabulated functions. We know that approximation of a Taylor based function requires very important condition that is the function should be in a closed form and not tabulated.

Another important fact to remember is that the Taylor based approximation is accurate only at the point of expansion and away from it the error appears to increase gradually.

Therefore, we will turn over to another topic of approximation called Lagrange due to its higher accuracy as compare to Taylor and also due to its flexibility to approximate tabulated functions.

To simplify the concept of Lagrange approximation [19], lets assume that we want to find a polynomial passing through two given points (x_0, y_0) and (x_1, y_1).

Assume that the function satisfying these two readings satisfies the following two equations:

$$f(x_0) = y_0 \tag{1}$$

$$f(x_1) = y_1 \tag{2}$$

Assume that a polynomial of first order takes the following form:

$$p(x) = \left(\frac{x - x_1}{x_0 - x_1}\right)y_0 + \left(\frac{x - x_0}{x_1 - x_0}\right)y_1 \tag{3}$$

Note that, the polynomial given by equation (3) will satisfy the following conditions:

When $x = x_0$

$$p(x_0) = y_0 = f(x_0) \tag{4}$$

And when $x = x_0$

$$p(x_1) = y_1 = f(x_1) \tag{5}$$

Now, let's again have at equation (3) and assume that the coefficients of both y_0 and y_1 will take other names as follows:

$$\ell_0(x) = \frac{x - x_1}{x_0 - x_1} \tag{6}$$

$$\ell_1(x) = \frac{x - x_0}{x_1 - x_0} \tag{7}$$

At the point $x = x_0$, the new coefficients will be:

$$\ell_0(x) = 1 \tag{8}$$

$$\ell_1(x) = 0 \tag{9}$$

Also the same coefficients at the other point $x = x_1$, will be:

$$\ell_0(x) = 0 \tag{10}$$

$$\ell_1(x) = 1 \tag{11}$$

Now, after some mathematical manipulations, one can get the general formula of Lagrange approximation as follows:

$$\ell(x)_{n,k} = \frac{(x - x_0)(x - x_1)(x - x_2)...(x - x_{k-1})..(x - x_n)}{(x_k - x_0)(x_k - x_1)(x_k - x_2)...(x_k - x_{k-1})..(x_k - x_n)} \tag{12}$$

A simplified form for equation (12) is as follows:

$$\ell(x)_{n,k} = \prod_{\substack{i=0 \\ i \neq k}}^{n} \frac{(x - x_i)}{(x_k - x_i)} \tag{13}$$

Now Lagrange polynomial can take the following form:

$$p(x) = f(x_0)\ell_{n,0}(x) + f(x_1)\ell_{n,1}(x) + \ldots\ldots + f(x_n)\ell_{n,n}(x) \tag{14}$$

Equation (14) in a compact form is given as:

$$p(x) = \sum_{k=0}^{n} f(x_k)\ell_{n,k}(x) \tag{15}$$

Where

$$\ell(x)_{n,k} = \prod_{\substack{i=0 \\ i \neq k}}^{n} \frac{(x - x_i)}{(x_k - x_i)} \tag{16}$$

Example

Use the data given in Table (**4**) below, to find a second degree polynomial for $f(x) = 1/x$

Table 4. Numerical data.

x_0	x_1	x_2
2.0	2.5	4.0
$f(x_0) = 0.5$	$f(x_1) = 0.4$	$f(x_2) = 0.25$

Solution

It is important to remind the reader a very important point while dealing with Lagrange approximation, that is, when dealing with tabulated functions, the number of terms assumed in the Lagrange polynomials will be equal to the number of points given.

Some Important Notations

1-Lagrange coefficient corresponding to x_0 is denoted as $\ell(x)_0$ and the same for all other points.

2-Each coefficient is a fraction consisting of a numerator and a denominator and both consist of a number of multiplied brackets; their number equal the number of given points minus one.

3-Each bracket in the numerator starts by the unknown variable x.

4-Each bracket in the denominator starts with the value of x in the table that corresponds to the Lagrange coefficient.

$$\ell_0(x) = \frac{(x - x_1)(x - x_2)}{(x_0 - x_1)(x_0 - x_2)}$$

$$\ell_1(x) = \frac{(x - x_0)(x - x_2)}{(x_1 - x_0)(x_1 - x_2)}$$

$$\ell_2(x) = \frac{(x - x_0)(x - x_1)}{(x_2 - x_0)(x_2 - x_1)}$$

- ***The coefficient corresponding to*** x_0

$$\ell_0(x) = \frac{(x - 2.5)(x - 4)}{(2.0 - 2.5)(2.0 - 4)}$$
$$= x^2 - 6.5x + 10$$

- ***The coefficient corresponding to*** x_1

$$\ell_1(x) = \frac{(x - 2.0)(x - 4)}{(2.5 - 2.0)(2.5 - 4)}$$
$$= -1.333x^2 + 8x - 10.667$$

- ***The coefficient corresponding to*** x_2

$$\ell_2(x) = \frac{(x - 2.0)(x - 2.5)}{(4.0 - 2.0)(4.0 - 2.5)}$$
$$= 0.333x^2 - 1.5x + 1.667$$

Substituting into Lagrange polynomial, leads to:

$$p(x) = f(x_0)\ell_0(x) + f(x_1)\ell_1(x) + f(x_2)\ell_2(x)$$

$$p(x) = (0.5)\bullet(x^2 - 6.5x + 10) + (0.4)\bullet(-1.333x^2 + 8x - 10.667)$$
$$+ (0.25)\bullet(0.333x^2 - 1.5x + 1.667)$$

Simplify, leads to:

$$p(x) = (0.05x - 0.425)x + 1.15$$

7. DIFFERENCE TABLES

Different tables represent measurements from the laboratory or field and basically constructed so as one can use them to interpolate or extrapolate missing data within the data or outside it.

May or may not the difference between two successive values of the independent variable is equal or not, which usually named as equal or un-equal spacing difference tables.

There are three types of difference tables namely, forward, backward and central difference tables. In the following sub-sections, we will introduce these types from the point how to construct and also how to use them.

Error propagation will be introduced and we will show how to decrease or stop this propagation [20].

7.1. Forward Differences

The forward difference basically denoted by the symbol Δ, is usually used when we have equal spacing data, *i.e.*, when the difference between two successive values of the independent variable is constant and this difference is denoted by $h = \Delta x$.

The number of rows in the tables is equal to the number of the given numerical data, while the number of columns depends on two different criteria, they are:

1-If all the values in any column become constant.

2-If the required certain accuracy to certain number of decimal places is achieved.

Differences of 1st order

Assume that we have n-tabulated values as follows:

x	x_1	x_2	....	x_n
$f(x)$	f_1	f_2	....	f_n

Then the 1st order differences can be written in the following manner:

$$\Delta f_1 = f_2 - f_1$$

$$\Delta f_2 = f_3 - f_2$$

$$\Delta f_3 = f_4 - f_3$$

$$\Delta f_i = f_{i+1} - f_i$$

Important Notes:

1-The first differences are written in the 3rd column, the second differences are written in the 4th column, and so on.

2-We know that $\Delta f_1 = f_2 - f_1$, this result is written in front of f_1, and so on.

The second order differences can be written as:

$$\Delta^2 f_1 = \Delta(\Delta f_1)$$
$$= \Delta(f_2 - f_1)$$
$$= \Delta f_2 - \Delta f_1$$
$$= (f_3 - f_2) - (f_2 - f_1)$$
$$= f_3 - 2f_2 + f_1$$

Similarly, the third order differences:

$$\Delta^3 f_1 = f_4 - 3f_3 + 3f_3 - f_1$$

In a general manner:

$$\Delta^n f_i = f_{i+n} - n f_{i+n-1} + \frac{n(n-1)}{2!} f_{i+n-2} \cdots$$

General Notations

When constructing the forward difference table, let us follow up the following notations:

1-Choose any point of the independent variable x and denote it as x_0, and so the corresponding value of the dependent variable y will be f_0

2-The values **above** f_0 will be numbered as f_{-1}, f_{-2},..., and so on.

3-The values **below** f_0 will be numbered as f_{+1}, f_{+2},..., and so on.

The following table is a forward difference table in a notation manner just to show how the construction of the Table (5) can occur.

Table 5. Symbolic forward difference table.

x	$f(x)$	$\Delta f(x)$	$\Delta^2 f(x)$	$\Delta^3 f(x)$	$\Delta^4 f(x)$
x_{-2}	f_{-2}				
x_{-1}	f_{-1}	Δf_{-2}			
x_0	f_0	Δf_{-1}	$\Delta^2 f_{-2}$		
x_1	f_1	Δf_0	$\Delta^2 f_{-1}$	$\Delta^3 f_{-2}$	
x_2	f_2	Δf_1	$\Delta^2 f_0$	$\Delta^3 f_{-1}$	$\Delta^4 f_{-2}$
x_3	f_3	Δf_2	$\Delta^2 f_1$	$\Delta^3 f_0$	$\Delta^4 f_{-1}$
x_4	f_4	Δf_3	$\Delta^2 f_2$	$\Delta^3 f_1$	$\Delta^4 f_0$

The following is a numerical forward difference Table (**6**), just as a numerical example.

Table 6. Numeric forward difference table.

x	$f(x)$	$\Delta f(x)$	$\Delta^2 f(x)$	$\Delta^3 f(x)$	$\Delta^4 f(x)$
0.0	0.000				
0.2	0.203	0.203			
0.4	0.423	0.220	0.017		
0.6	0.684	0.261	0.041	0.024	
0.8	1.030	0.346	0.025	0.044	0.020
1.0	1.557	0.527	0.181	0.096	0.052
1.2	2.572	1.015	0.488	0.307	0.211

7.2. Backward Differences

The backward difference basically denoted by the symbol ∇ is usually used when we have equal spacing data, *i.e.*, when the difference between two successive values of the independent variable is constant and this difference is denoted by $h = \Delta x$.

The number of rows in the tables is equal to the number of the given numerical data, while the number of columns depends on two different criteria, they are:

1-If all the values in any column become constant.

2-If the required accuracy to certain number of decimal places is achieved.

Differences of 1st order

Assume that we have n-tabulated values as follows:

x	x_1	x_2	....	x_n
$f(x)$	f_1	f_2	....	f_n

Then the 1storder differences can be written in the following manner:

$$\nabla f_1 = f_1 - f_0$$

$$\nabla f_2 = f_2 - f_1$$

$$\nabla f_3 = f_3 - f_2$$

$$\nabla f_i = f_i - f_{i-1}$$

Important Notes:

1-The first differences are written in the 3rd column, the second differences are written in the 4thcolumn, and so on.

2-We know that $\nabla f_1 = f_1 - f_0$, this result is written in front of f_1, and so on.

Backward differences of second order are as follows:

$$\begin{aligned}
\nabla^2 f_1 &= \nabla(\nabla f_1) \\
&= \nabla(f_1 - f_0) \\
&= \nabla f_1 - \Delta f_0 \\
&= (f_1 - f_0) - (f_0 - f_{-1}) \\
&= f_1 - 2f_0 + f_{-1}
\end{aligned}$$

In general

$$\nabla^2 f_i = f_i - 2f_{i-1} + f_{i-2}$$

Table 7. Symbolic backward difference table.

x	$f(x)$	$\nabla f(x)$	$\nabla^2 f(x)$	$\nabla^3 f(x)$	$\nabla^4 f(x)$
x_{-2}	f_{-2}	∇f_{-1}	$\nabla^2 f_0$	$\nabla^3 f_1$	$\nabla^4 f_2$
x_{-1}	f_{-1}				$\nabla^4 f_3$

Table 7 contd….

x_0	f_0				$\nabla^4 f_4$
x_1	f_1			$\nabla^3 f_4$	
x_2	f_2		$\nabla^2 f_4$		
x_3	f_3	∇f_4			
x_4	f_4				

The following is a numerical backward difference table, just as a numerical example.

Table 8. Numeric forward difference table.

x	$f(x)$	$\nabla f(x)$	$\nabla^2 f(x)$	$\nabla^3 f(x)$	$\nabla^4 f(x)$
0.0	0.000	0.203	0.017	0.024	0.020
0.2	0.203				
0.4	0.423				0.211
0.6	0.684			0.307	
0.8	1.030		0.488		
1.0	1.557	1.015			
1.2	2.572				

7.3. Divide Differences

This is a general difference table since it can be used for equal and unequal space data tables. It is denoted by δ, and has special criteria during construction. The details of construction of such difference Table (**9**) are shown in the next table.

Table 9. Symbolic central difference table.

x	y	δ : First Approximation	δ : Second Approximation
x_0	y_0		
		$\delta(x_1, x_0) = \dfrac{y_1 - y_0}{x_1 - x_0}$	
x_1	y_1		$\delta(x_2, x_1, x_0) = \dfrac{\delta(x_2, x_1) - \delta(x_1, x_0)}{x_2 - x_0}$
		$\delta(x_2, x_1) = \dfrac{y_2 - y_1}{x_2 - x_1}$	
x_2	y_2		
		$\delta(x_3, x_2) = \dfrac{y_3 - y_2}{x_3 - x_2}$	$\delta(x_3, x_2, x_1) = \dfrac{\delta(x_3, x_2) - \delta(x_2, x_1)}{x_3 - x_1}$
x_3	y_3		
		$\delta(x_4, x_3) = \dfrac{y_4 - y_3}{x_4 - x_3}$	
x_4	y_4		$\delta(x_4, x_3, x_2) = \dfrac{\delta(x_4, x_3) - \delta(x_3, x_2)}{x_4 - x_2}$
		$\delta(x_5, x_4) = \dfrac{y_5 - y_4}{x_5 - x_4}$	
x_5	y_5		$\delta(x_5, x_4, x_3) = \dfrac{\delta(x_5, x_4) - \delta(x_4, x_3)}{x_5 - x_3}$

8. FIRST DERIVATIVE FROM FORWARD INTERPOLATION

We sometimes encounter difficulties when measuring or collecting data, *e.g.*, the survey engineer encounters difficulty in taking real measurements.

If he encounters lake of water in the area he works, he will need interpolation or extrapolation techniques to get the measurements he already could not get.

The basic idea is that, we derive a polynomial having the same criteria of the same original function. We know from the approximation theory that approximation of function by a specified polynomial has the property that both the slope of function and the polynomial at the point of approximation are nearly the same, but going away from that point, leading to an increase of accumulated error.

Let us start derivation, by writing Newton-Gregory for forward interpolation as follows [21-31]:

$$f(x_S) = P_n(x_S) + error$$

$$= f_0 + S\Delta f_0 + \binom{S}{2}\Delta^2 f_0 + \dots + \binom{S}{n}\Delta^n f_0 + error \qquad (1)$$

The second term in the right hand-side of equation (1) can be evaluated from the following formula:

$$error\ of\ P_n(x_S) = \binom{S}{n+1}h^{n+1}f^{(n+1)}(\xi),\ \ x_0 < \xi < x_n \qquad (2)$$

Now differentiating equation (1) and by taking into consideration that both f_0 and all terms containing Δ are constant, then one can obtain the following equation:

$$f'(x_S) \approx P'_n(x_S) = \frac{d}{ds}\big(P_n(x_S)\big)\frac{ds}{dx} = \frac{d}{ds}\big(P_n(x_S)\big)\frac{1}{h} \qquad (3)$$

$$f'(x_S) \approx P'_n(x_S)$$
$$= \frac{1}{h}\left(\Delta f_0 - \frac{1}{2}\Delta^2 f_0 + \frac{1}{6}\left(\frac{(s-1)(s-2)}{+s(s-2)+s(s-1)}\right)\Delta^3 f_0 + \dots\right) \qquad (4)$$

Put $s = 0$ into equation (4), leads to:

$$f'(x_0) \approx \frac{1}{h}\left(\Delta f_0 - \frac{1}{2}\Delta^2 f_0 + \frac{1}{3}\Delta^3 f_0 - \frac{1}{4}\Delta^4 f_0 \dots \pm \frac{1}{n}\Delta^n f_0\right) \qquad (5)$$

Now if we have an interpolation table, one can easily and by making use of equation (5), evaluate the first derivative of the function at any specified point within the limits of the data of the table.

$$error\ of\ P'_n(x_0) = h^{n+1}f^{(n+1)}(\xi)\left[\frac{(-1)^n n!}{(n+1)!}\right]\left(\frac{1}{h}\right)$$

$$= \frac{(-1)^n}{n+1}h^n f^{(n+1)}(\xi)$$

Example

Use the data in table to estimate the first derivative of y at $x = 1.7$, using $h = 0.2$ and compute using one, two, three or four terms of the first derivative formula.

Solution

Here we have equal space data and it is required to find the first derivative using three different cases. As we can see, the point at which it is required to find the derivative is also a point in the given data.

Therefore, we will use $x = 1.7$ as a reference point for further computations. Also, because we have equal space data, we can use forward, backward or central difference table.

Another important reminder is that it is not necessary, as we said before that in case of forward difference, to write the difference in front of the first value in each row, but we can write down the difference in between the two successive values, and this is what we will do in the present example.

Table 10. Numerical data with forward difference table.

x	y	Δy	$\Delta^2 y$	$\Delta^3 y$	$\Delta^4 y$
1.3	3.669				
		0.813			
1.5	4.482		0.179		
		0.992		0.041	
1.7	5.474		0.220		0.007
		1.212		0.048	
1.9	6.686		0.268		0.012
		1.480		0.060	
2.1	8.166		0.328		0.012
		1.808		0.072	
2.3	9.974		0.400		
		2.208			
2.5	12.182				

The given data is for the function $y = e^x$ so that we can compare the numerical results with the exact result.

Now, let us start with one term solution:

With **one** term

$$y'(1.7) = \frac{1}{0.2}(1.212) = 6.060$$

With **two** terms

$$y'(1.7) = \frac{1}{0.2}\left[(1.212) - \frac{1}{2}(0.268)\right] = 5.390$$

With **three** terms

$$y'(1.7) = \frac{1}{0.2}\left[(1.212) - \frac{1}{2}(0.268) + \frac{1}{3}(0.060)\right] = 5.490$$

With **four** terms

$$y'(1.7) = \frac{1}{0.2}\left[(1.212) - \frac{1}{2}(0.268) + \frac{1}{3}(0.060) - \frac{1}{4}(0.012)\right] = 5.475$$

The exact numerical value of the derivative at the indicated point is $y'(1.7) = e^{1.7} = 5.474$, while the computed value to four terms is 5.475. If we apply error equation to find error associated with the evaluation of the first derivative, it will be as follows:

$$error = \frac{(-1)^n}{n+1}h^n f^{(n+1)}(\xi)$$

With **one** term:

$$error = \frac{(-1)^1}{1+1}h^1 f^{(2)}(\xi), \quad 1.7 \le \xi \le (1.7 + h) = 1.9$$

$$error = \frac{(-1)}{2}(0.2)^1 \left\{ \begin{matrix} e^{1.7} & min \\ e^{1.9} & max \end{matrix} \right\} = \left\{ \begin{matrix} -0.547 & min \\ -0.669 & max \end{matrix} \right\}$$

With *two* terms:

$$error = \frac{(-1)^{21}}{2+1} h^2 f^{(3)}(\xi) , \quad 1.7 \le \xi \le (1.7 + 2h) = 2.1$$

$$error = \frac{1}{3}(0.04) \begin{Bmatrix} e^{1.7} & \min \\ e^{2.1} & \max \end{Bmatrix} = \begin{Bmatrix} 0.073 & \min \\ 0.109 & \max \end{Bmatrix}$$

With *three* terms:

$$error = \frac{(-1)^{31}}{3+1} h^3 f^{(4)}(\xi) , \quad 1.7 \le \xi \le (1.7 + 3h) = 2.3$$

$$error = \frac{-1}{4}(0.008) \begin{Bmatrix} e^{1.7} & \min \\ e^{2.3} & \max \end{Bmatrix} = \begin{Bmatrix} -0.011 & \min \\ -0.020 & \max \end{Bmatrix}$$

With *four* terms:

$$error = \frac{(-1)^{41}}{4+1} h^4 f^{(5)}(\xi) , \quad 1.7 \le \xi \le (1.7 + 4h) = 2.5$$

$$error = \frac{1}{5}(0.0016) \begin{Bmatrix} e^{1.7} & \min \\ e^{2.5} & \max \end{Bmatrix} = \begin{Bmatrix} 0.002 & \min \\ 0.004 & \max \end{Bmatrix}$$

9. HIGHER DERIVATIVES

The higher derivatives formulas can be derived from differentiation of the formula of the first order derivative. Another point should be taken into consideration is that to find the error formula corresponding to each derivative, put $s = 0$ in the corresponding derivative formula.

Let us consider an example to show how the second order derivative formula is obtained, and the final result will be as follows:

$$f''(x_0) \cong \frac{1}{h^2}\left(\Delta^2 f_0 - \Delta^3 f_0 + \frac{11}{12} \Delta^4 f_0 - \frac{5}{6} \Delta^5 f_0 + ... \right)$$

Example

Use the data in table to estimate the second derivative of y at $x = 1.7$, use $h = 0.2$.

Solution

$$f''(x_0) \cong \frac{1}{h^2}\left(\Delta^2 f_0 - \Delta^3 f_0 + \frac{11}{12}\Delta^4 f_0 - \frac{5}{6}\Delta^5 f_0 + ... \right)$$

$$f''(x_0) \cong \frac{1}{(0.2)^2}\left(0.268 - 0.06 + \frac{11}{12}(0.012) \right)$$
$$= 5.475$$

Table 11. Numerical data with forward difference table.

x	y	Δy	$\Delta^2 y$	$\Delta^3 y$	$\Delta^4 y$
1.3	3.669				
		0.813			
1.5	4.482		0.179		
		0.992		0.041	
1.7	5.474		0.220		0.007
		1.212		0.048	
1.9	6.686		$\Delta^2 y_0=0.268$		0.012
		1.480		0.060	
2.1	8.166		0.328		0.012
		1.808		0.072	
2.3	9.974		0.400		
		2.208			
2.5	12.182				

10. LOZENGE DIAGRAM FOR DERIVATIVES

Lozenge diagram is a tabulated method for the derivatives of a function at specified points. Lozenge diagram has the property that it is easy to construct [32]. The diagram can be considered in two parts:

(1)The differences of the function;

(2)The interspersed coefficients, one superimposed on the other.

Table 12. Lozenge diagram for derivatives.

f_{-3}	1	$\Delta^2 f_{-4}$	13/3	$\Delta^4 f_{-5}$	137/60
0	Δf_{-3}	5/2	$\Delta^3 f_{-4}$	25/12	$\Delta^5 f_{-5}$
f_{-2}	1	$\Delta^2 f_{-3}$	11/6	$\Delta^4 f_{-4}$	1/5
0	Δf_{-2}	3/2	$\Delta^3 f_{-3}$	1/4	$\Delta^5 f_{-4}$
f_{-1}	1	$\Delta^2 f_{-2}$	1/3	$\Delta^4 f_{-3}$	-1/20
0	Δf_{-1}	1/2	$\Delta^3 f_{-2}$	-1/12	$\Delta^5 f_{-3}$
f_0	1	$\Delta^2 f_{-1}$	-1/6	$\Delta^4 f_{-2}$	1/30
0	Δf_{-1}	-1/2	$\Delta^3 f_{-1}$	1/12	$\Delta^5 f_{-2}$
f_1	1	$\Delta^2 f_0$	1/3	$\Delta^4 f_{-1}$	-1/20
0	Δf_{-2}	-3/2	$\Delta^3 f_0$	-1/4	$\Delta^5 f_{-1}$
f_2	1	$\Delta^2 f_1$	11/6	$\Delta^4 f_0$	1/5
0	Δf_{-3}	-5/2	$\Delta^3 f_1$	-25/12	$\Delta^5 f_0$
f_3	1	$\Delta^2 f_2$	13/3	$\Delta^4 f_1$	137/60

Consider the array of coefficients in the above figure:

	1		13/3		137/60
0		5/2		25/12	
	1		11/6		1/5
0		3/2		1/4	
	1		1/3		-1/20
0		1/2		-1/12	
	1		-1/6		1/30
0		-1/2		1/12	
	1		1/3		-1/20
0		-3/2		-1/4	
	1		11/6		1/5
0		-5/2		-25/12	
	1		13/3		137/60

To understand how to deal with Lozenge diagram, let's consider a horizontal path as follows:

f_{-3}	0	$\Delta^2 f_{-4}$	3	$\Delta^4 f_{-5}$	15/4
0	Δf_{-3}	1	$\Delta^3 f_{-4}$	35/12	$\Delta^5 f_{-5}$
f_{-2}	0	$\Delta^2 f_{-3}$	2	$\Delta^4 f_{-4}$	5/6
0	Δf_{-2}	1	$\Delta^3 f_{-3}$	11/12	$\Delta^5 f_{-4}$

f_{-1}	0	$\Delta^2 f_{-2}$	1	$\Delta^4 f_{-3}$	-1/12
0	Δf_{-1}	1	$\Delta^3 f_{-2}$	$-1/12$	$\Delta^5 f_{-3}$
f_0	0	$\Delta^2 f_{-1}$	0	$\Delta^4 f_{-2}$	0
0	Δf_0	1	$\Delta^3 f_{-1}$	-1/12	$\Delta^5 f_{-2}$
f_1	0	$\Delta^2 f_0$	-1	$\Delta^4 f_{-1}$	1/12
0	Δf_1	1	$\Delta^3 f_0$	11/12	$\Delta^5 f_{-1}$
f_2	0	$\Delta^2 f_1$	-2	$\Delta^4 f_0$	-5/6
0	Δf_2	1	$\Delta^3 f_1$	35/12	$\Delta^5 f_0$
f_3	0	$\Delta^2 f_2$	-3	$\Delta^4 f_1$	-15/4

Beginning at f_0:

$$\frac{df}{dx}\bigg|_{x=x_o} = \frac{1}{h}\left[\frac{\Delta f_0 + \Delta f_{-1}}{2} + (0)\Delta^2 f_{-1} + \left(-\frac{1}{6}\right)\frac{\Delta^3 f_{-1} + \Delta^3 f_{-2}}{2} + ...\right]$$

Or

$$\frac{df}{dx}\bigg|_{x=x_o} = \frac{1}{h}\left[\frac{(f_1 - f_0) + (f_0 - f_{-1})}{2} + 0(h^3)\right] = \left[\frac{f_1 - f_{-1}}{2h}\right] + 0(h^2)$$

$$\frac{d^2 f}{dx^2}\bigg|_{x=x_o} = \frac{1}{h^2}\left[\Delta^2 f_{-1} + 0 + \left(-\frac{1}{12}\right)\Delta^4 f_{-2} + ...\right]$$

Or

$$\frac{d^2 f}{dx^2}\bigg|_{x=x_o} = \left[\frac{f_1 - 2f_0 + f_{-1}}{h^2}\right] + 0(h^2)$$

Example

Estimate the second derivative of y at $x = 1.7$, use $h = 0.2$

Solution

Construct the forward difference table as follows, see Table **13**:

Table 13. Data and forward difference table.

x	y	Δy	$\Delta^2 y$	$\Delta^3 y$	$\Delta^4 y$
1.3	3.669				
		0.813			
1.5	4.482		0.179		
		0.992		0.041	
1.7	5.474		0.220		0.007
		1.212		0.048	
1.9	6.686		0.268		0.012
		1.480		0.060	
2.1	8.166		0.328		0.012
		1.808		0.072	
2.3	9.974		0.400		
		2.208			
2.5	12.182				

$$\frac{d^2 f}{dx^2}\bigg|_{x=x_o} = \left[\frac{f_1 - 2f_0 + f_{-1}}{h^2} \right] + 0(h^2)$$

$$\left.\frac{d^2 f}{dx^2}\right|_{x=1.7} = \left[\frac{6.686 - 2(5.474) + 4.482}{(0.2)^2}\right]$$
$$= \frac{0.220}{(0.2)^2}$$
$$= 5.5$$

11. NUMERICAL INTEGRATION

Mathematics is a language, and unfortunately, despite this language is available in the mind of all humans, only some of them have the ability to understand it.

Till now and up to the end of our life, there will be secrets for us, but trials exist. Suppose that one asks to find exact form of solution for:

$$\int_a^b e^2 dx$$

It seems simple but in fact it is still so difficult to find such a close solution. From that point and other points, researchers think about numerical approaches, especially with the rapid development in computers and advanced programming.

Numerical integration is one of the most important topics in mathematics and going up with the recent numerical techniques, such as finite element method, boundary elements methods and more mesh-free methods.

In next sections, we will try to introduce the basic and classical numerical integration. Definite integral has a general form as follows (see Fig. **1**):

$$I(f) = \int_a^b f(x)dx = F(b) - F(a) \tag{1}$$

In equation (1), the function $F(x)$ represents the final closed form of solution of the integral.

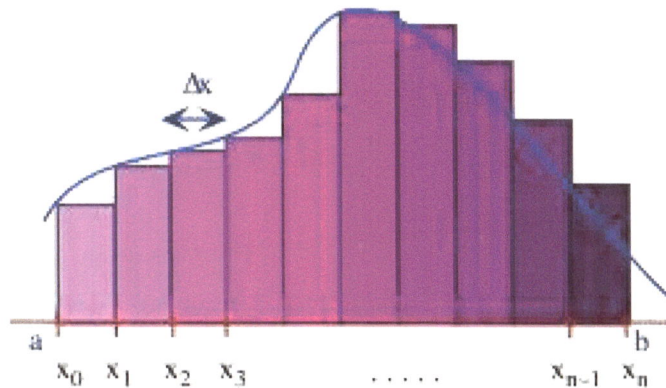

Fig. (1). Discretization the area under the plane curve.

12. NEWTON-COTES INTEGRATION FORMULAS

The usual strategy in developing formulas for numerical integration is to pass a polynomial through points of the function, and then integrate this polynomial approximation to the function.

This permits us to integrate a function known only as a table of values. When the values of the function are equally spaced, Newton-Gregory forward polynomial [33] is a convenient starting point, therefore,

$$I(f) = \int_a^b f(x)dx \cong \int_a^b P_n(x_s)dx \tag{2}$$

As we know in the numerical analysis, there are no exact values at all and usually there is some error due to the integral in equation (2), and that error can be evaluated from the following equation:

$$Error = \int_a^b \left(\frac{s}{n+1}\right) h^{n+1} f^{(n+1)}(\xi)dx \tag{3}$$

Now, let us start developing the Newton-Cotes formula of integration, and taking into consideration that we will replace the variable x to another variable s and so $dx = hds$

13. TRAPEZOIDAL RULE

The basic idea of trapezoidal rule for integration is to divide the area under curve into a sufficient number of vertical trapezoidal strips, then find Reimman sum for these strips and take the limit when the number of such strips tends to infinity.

13.1. Trapezoid Area

The area of a trapezoid is the average length of the parallel sides, times the distance between them. Assume that we have a function $f(x)$ defined over a closed interval [a, b], as shown in Fig. (2).

Divide the area under the function curve to an infinite number of vertical strips, each of which width Δx. Suppose that, we want to evaluate the line integral over the interval $x_0 \to x_1$, therefore, the line integral can take the following form:

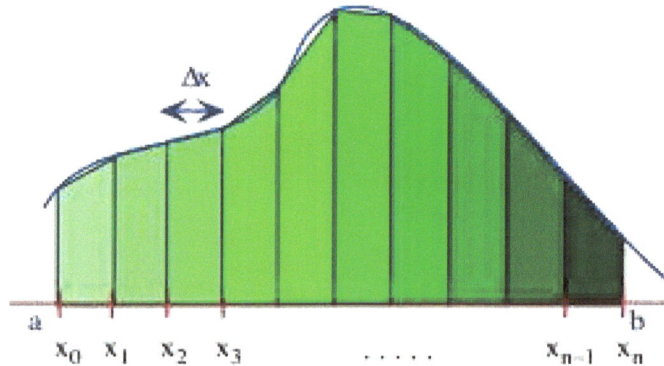

Fig. (2). Trapezoidal rule.

$$\int_{x_o}^{x_1} f(x)dx \cong \int_{x_o}^{x_1} (f_o + s\Delta f_o)dx \tag{4}$$

The right hand side of the equation (4), in terms of the new variables, can take the following form:

$$\int_{x_o}^{x_1} (f_o + s\Delta f_o)dx = h \int_{s=0}^{s=1} (f_o + s\Delta f_o)ds = \frac{h}{2}(f_o + f_1) \tag{5}$$

From equations (4) and (5), one can get the following form for integration over the interval $x_0 \rightarrow x_1$:

$$\int_{x_o}^{x_1} f(x)dx \cong \frac{h}{2}(f_o + f_1) \tag{6}$$

Basically as we said before, there will be some error, referring to Fig. (3) below. If we approximate the part of the curve to straight line, then the error appears, and this error can then be evaluated from the following formula:

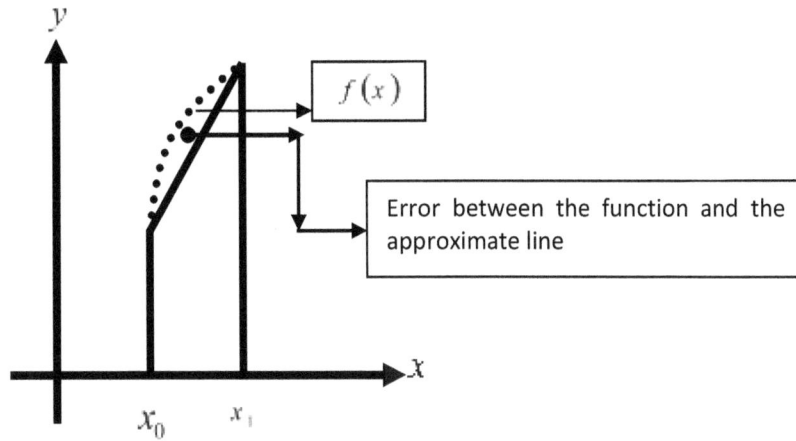

Fig. (3). Approximation of one strip.

$$Error = -\frac{1}{12}h^3 f''(\xi_1) \quad , x_o \leq \xi_1 \leq x_1 \tag{7}$$

A generalized formula for integration between two successive points can be written as:

$$I(f) = \int_a^b f(x)dx \cong \left\{ \frac{f(x_i) + f(x_{i+1})}{2} \right\} \Delta x \tag{8}$$

If one assumes that:

$$\Delta x = h \tag{9}$$

Then equation (8) will be:

$$I(f) = \int_a^b f(x)dx \cong \frac{h}{2}\{f(x_i) + f(x_{i+1})\} \qquad (10)$$

Equation (10) can then be generalized as follows:

$$I(f) = \int_a^b f(x)dx \cong \frac{h}{2}\{f_1 + f_2 + f_2 + f_3 + f_3 + \ldots + f_n + f_{n+1}\} \qquad (11)$$

We see that on the right-hand side that each term repeats twice except the first and the last, therefore, we can re-write it again in a simple form as follows:

$$I(f) = \int_a^b f(x)dx \cong \frac{h}{2}\left\{f_1 + f_{n+1} + \sum_{i=2}^n f_i\right\} \qquad (12)$$

Equation (12) is the general formula for trapezoidal rule for evaluating the numerical integration.

Example

For the following tabulated function given below. Evaluate the integration of that function between 1.8 and 3.8.

Solution

Table 14. The numerical data function over the integration interval.

x	$f(x)$	x	$f(x)$
1.6	4.953	2.8	16.445
1.8	$f_1 = 6.05$	3.0	20.086
2.0	7.389	3.2	24.533
2.2	9.025	3.4	29.964
2.4	11.023	3.6	36.598
2.6	13.464	3.8	$f_{n+1} = 44.701$

The first step in the solution procedure is to write the trapezoidal formula for integration:

$$I(f) = \int_a^b f(x)dx \cong \frac{h}{2}\left\{f_1 + f_{n+1} + \sum_{i=2}^{n} f_i\right\}$$

Next, is to put the formula according to the required integration:

$$I(f) = \int_{1.8}^{3.8} f(x)dx$$

As shown in the table, the lower limit of integration is $x = 1.8$ and the upper limit will be $x = 3.8$. The first value of the function that corresponds to $x = 1.8$ is $f_1 = 6.05$ and the last value that corresponds to $x = 3.8$ is $f_{n+1} = 44.701$.

Now let us find the summation of all values that appear in the trapezoidal rule except the first and the last as follows:

$$f_2 + f_3 + \ldots\ldots + f_n = \left|\begin{array}{l} 7.389 + 9.025 + 11.023 + 13.464 + 16.445 \\ + 20.086 + 24.533 + 29.964 + 36.598 \end{array}\right|$$

Now let us substitute from the first, last and the summation of the middle terms in the trapezoidal formula, which leads to:

$$I(f) = \int_a^b f(x)dx \cong$$

$$\frac{0.2}{2}\left\{(6.05 + 44.701) + +2\bullet\left[\begin{array}{l} 7.389 + 9.025 + 11.023 + 13.464 + 16.445 \\ + 20.086 + 24.533 + 29.964 + 36.598 \end{array}\right]\right\}$$

$$= 23.9944$$

Example

Evaluate the following integral $I = \int_0^1 \left(\frac{1}{1+x}\right)dx$ using trapezoidal rule and number of sub-division equals 6.

Solution

In the present example, it is required to evaluate the numerical value of the given integral in the same time, no tabulated values are given. Therefore, the first step is to construct a numerical table, so as we can apply the trapezoidal rule for integration. The integrand is as follows:

$$f(x) = \left(\frac{1}{1+x} \right)$$

The table will range from $x = 0$ and $x = 1$, due to the limits of integration. Now to complete construction the numerical data, we should know the number of intervals, and this is already given in the example itself. The number of sub-divisions is equal to 6; therefore, the following empirical relation will be used to find step width:

$$h = \frac{b-a}{n} = \frac{1-0}{6} = \frac{1}{6}$$

Now, we have all data required to construct the table, and it is provided in Table **15**.

Table 15. The numerical data function over the integration interval.

x	0	1/6	2/6	3/6	4/6	5/6	1
$f(x)$	$f_1 = 1$	6/7	6/8	6/9	6/10	6/11	$f_7 = \dfrac{1}{2}$

As shown in the table, the lower limit of integration is $x = 0$ and the upper limit will be $x = 1$. The first value of the function that corresponds to $x = 0$ is $f_1 = 1$ and the last value that corresponds to $x = 1$ is $f_{n+1} = \dfrac{1}{2}$.

$$I(f) = \int_a^b f(x)dx \cong \frac{h}{2}\left\{f_1 + f_{n+1} + 2\sum_{i=2}^n f_i\right\} = \int_{a=0}^{b=1} f(x)dx$$

$$\cong \frac{1}{12}\left\{f_1 + f_7 + 2\sum_{i=2}^6 f_i\right\}$$

$$= \int_{a=0}^{b=1} f(x)dx \cong \frac{1}{12}\{f_1 + f_7 + 2(f_2 + f_3 + f_4 + f_5 + f_6)\}$$

$$I(f) = \int_a^b f(x)dx$$

$$\cong \frac{0.2}{2}\left\{\left(1 + \frac{1}{2}\right) + +2 \bullet \left[\frac{6}{7} + \frac{6}{8} + \frac{6}{9} + \frac{6}{10} + \frac{6}{11}\right]\right\}$$

$$= 0.6948$$

14. SIMPSON'S METHOD

14.1. Simpson's 1/3 Rule

When evaluating integrals, using trapezoidal rule, there will be error, due to the approximation of the function within each interval to a straight line and so there will be an area between the part of the function and the approximated straight line, see Fig. (**4**).

This error increases for finite number of sub-interval. In the present section, we will turn our attention to another and more accurate numerical method for evaluating the integral.

Applying Newton-Cotes formula from x_0 to x_2, leads to:

$$\int_{x_o}^{x_2} f(x)dx \cong \frac{h}{3}(f_o + 4f_1 + f_2) - \frac{h^5}{90}f^{(5)}(\xi) \tag{13}$$

The last term in equation (13) represents the error due to the approximation described above. Now if one generalizes equation (13) over the interval of integration from $x = a$ to $x = b$, this will lead to the following generalized formula of integration:

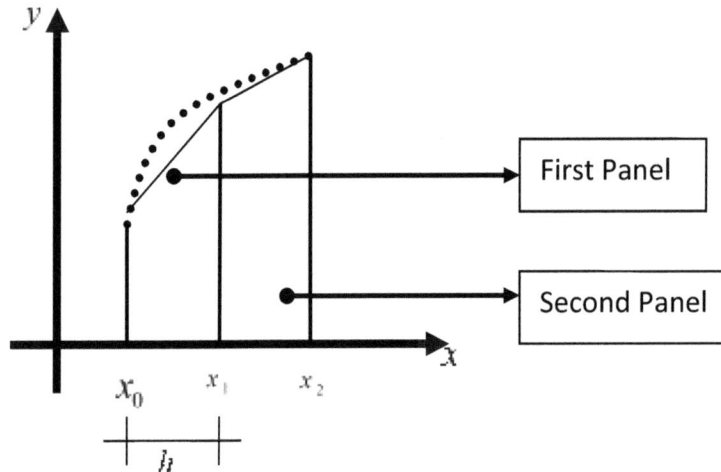

Fig. (4). Simpson'1/3rule.

$$\int_a^b f(x)dx \cong \frac{h}{3}\left(f_{\substack{starting \\ value}} + f_{\substack{end \\ value}} + 2\times\left[\sum_{i=odd} f_i\right] + 4\times\left[\sum_{i=even} f_i\right] \right) \quad (14)$$

Equation (14) is the Simpson'1/3rule for evaluation the integral over a closed interval $[a, b]$. As usual, there will be some error, and this error can be evaluated from the following formula:

$$Error = -\left(\frac{b-a}{180}\right)h^4 f^{(4)}(\xi_1) \quad, a \le \xi_1 \le b \quad (15)$$

*There is a very important rule when using Simpson'1/3 rule, that is, the number of intervals must be an **even number**.*

Example

Evaluate the integration of the tabulated function below between 1.8 and 3.4, see Table (**16**).

Table 16. Data function.

x	$f(x)$	x	$f(x)$
1.6	4.953	2.8	16.445
1.8	$f_s = 6.05$	3.0	20.086
2.0	7.389	3.2	24.533
2.2	9.025	3.4	$f_e = 29.964$
2.4	11.023	3.6	36.598
2.6	13.464	3.8	44.701

Solution

This example is the same solved before using trapezoidal. Just to see the accuracy between the trapezoidal and Simpson'1/3 rules, let's start the solution procedure by re-writing the basic formula that will be used in the computation:

$$\int_a^b f(x)dx \cong \frac{h}{3}\left(f_{\substack{starting \\ value}} + f_{\substack{end \\ value}} + 2\times\left[\sum_{i=odd} f_i\right] + 4\times\left[\sum_{i=even} f_i\right] \right)$$

Now, follow the same procedure described before. According to the formula:

$$\int_a^b f(x)dx \cong \left(\frac{0.2}{3}\right)\left(\begin{array}{l} 6.05+29.964+2\times[9.025+13.464+20.086] \\ +4\times[7.389+11.023+16.445+24.533] \end{array} \right)$$

$$= 23.915$$

14.2. Simpson's (3/8) Rule

Usually, we need the best and if this is a common property for all people, scientists have this property in their research work.

As mentioned before, there is usually some error between the methods for the same purpose, due to that, we will introduce Simpson's 3/8 rule which is more accurate than Simpson'1/3 rule.

Following the same procedure for developing Simpson'1/3 rule, one can get the following formula for evaluating the integral numerically:

$$\int_{x_o}^{x_3} f(x)dx \cong \frac{3h}{8}\left(f_o + 3f_1 + 3f_2 + f_3\right) \tag{16}$$

The error in equation (16) can be evaluated from the following formula:

$$Error = -\frac{3}{80}h^5 f^{(5)}(\xi_1) \quad , x_o \le \xi_1 \le x_3 \tag{17}$$

To generalize the formula given by equation (1), let us follow the following:

$$h = \frac{b-a}{n}, \quad x_i = a + ih$$

Then, we have:

$$\int_a^b f(x)dx \cong \frac{3h}{8}\left(\underbrace{f_o}_{\text{Initial}} + \underbrace{f_n}_{\text{Final}} + \underbrace{3f_1 + 3f_2}_{\text{Coeff}=3} + 2f_3 + \underbrace{3f_4 + 3f_5}_{\text{Coeff}=3} + 2f_6 + ... + \underbrace{3f_{n-3} + 3f_{n-2}}_{\text{Coeff}=3} + 2f_{n-1} \right)$$

It is very important to note in this method that, we can only use it if n is a multiple of three.

15. ROMBERG'S INTEGRATION METHOD

This method is more accurate than Newton-Cotes formulas [34-38]. Its implementation starts by assuming the area of a trapezoid under the whole area. It is used to evaluate an integral of the form:

$$\int_a^b f(x)dx$$

by applying Richardson extrapolation (Richardson 1911) repeatedly on the trapezium rule or the rectangle rule (midpoint rule). The estimates generate a triangular array. Romberg's method is a Newton–Cotes formula – it evaluates the integrand at equally spaced points. The integrand must have continuous derivatives, though fairly good results may be obtained if only a few derivatives exist. If it is possible to evaluate the integrand at unequally spaced points, then other methods such as Gaussian quadrature and Clenshaw–Curtis quadrature are generally more accurate.

15.1. Solution Procedure

1-Let the area under the function be a trapezoid ***once a time***, and find the value of the integration, from:

$$I_{1,1} = \frac{h}{2}(f_0 + f_1)$$

2-Divide the area under the function into ***two*** areas, then evaluate the integration from:

$$I_{2,1} = \frac{h}{4}\left(f_0 + f_{\frac{1}{2}}\right) + \frac{h}{4}\left(f_{\frac{1}{2}} + f_1\right)$$

$$= \frac{h}{4}(f_0 + f_1) + \frac{h}{2}\left(f_{\frac{1}{2}1}\right)$$

$$= \frac{1}{2}\left(I_{1,1} + hf_{\frac{1}{2}}\right)$$

3-Divide the area under the function into ***three*** areas, then evaluate the integration from:

$$I_{3,1} = \frac{1}{2}\left(I_{2,1} + \frac{h}{2}\left(f_{\frac{1}{4}} + f_{\frac{3}{4}}\right)\right)$$

Note that if one writes any of the integral above as $I_{i,j}$ the subscript (i) denotes the level of approximation, while (j) denotes the iteration number.

Similar procedure, for the second iteration and starting from one, two and three subdivisions, one can get the following second iteration relations:

$$I_{1,2} = \frac{1}{3}\left(4I_{2,1} - I_{1,1}\right)$$

$$I_{2,2} = \frac{1}{3}\left(4I_{3,1} - I_{2,1}\right)$$

$$I_{3,2} = \frac{1}{3}\left(4I_{4,1} - I_{3,1}\right)$$

Therefore, Romberg general formula then takes the following form:

$$I_{i,j} = \frac{1}{\left(4^{j-1} - 1\right)}\left(4^{j-1}I_{i+1,j-1} - I_{i,j-1}\right) \qquad i = 1,2,3,\dots \; \& \; j = 2,3,4,\dots$$

Example

Using Romberg method to evaluate the integral:

$$\int_0^8 \left(\frac{5}{8}x^4 - 4x^3 + 2x + 1\right)dx$$

Solution

The first step is to construct the following table starting from the lower limit of integration to the upper limit of integration, see Table (**17**).

Table 17. Data function.

x	0	1	2	3	4
$f(x)$	1	-0.375	-17	-50.375	-87
x	5	6	7	8	
$f(x)$	-98.375	-41	143.625	529	

Now follow the procedure described above:

$$I_{1,1} = \frac{h}{2}(f_0 + f_1)$$

$$= \frac{8}{2}(1 + 529)$$

$$= 2120$$

$$I_{2,1} = \frac{1}{2}\left(I_{1,1} + hf_{\frac{1}{2}}\right)$$

$$= \frac{1}{2}(2120 + 8 \bullet (-87))$$

$$= 712$$

$$I_{3,1} = \frac{1}{2}\left(I_{2,1} + \frac{h}{2}\left(f_{\frac{1}{4}} + f_{\frac{3}{4}}\right)\right)$$

$$= \frac{1}{2}\left(712 + \frac{8}{2}(-17 - 41)\right)$$

$$= 240$$

To improve the accuracy of the results, construct the following Table (**18**).

Table 18. Romberg method.

$I_{i,1}$	$I_{i,2}$	$I_{i,3}$	$I_{i,4}$
2120	242.6667	72.00003	
712	82.6667	72.00003	
240	72.6667		
114.5			

Then the approximate value of the given integral will be 72.00003.

16. MULTIPLE INTEGRALS

In this section, we will learn how to evaluate numerically double integral having the following general form:

$$I = \int_{c}^{d} \int_{a}^{b} f(x,y)\,dx\,dy$$

The main idea for evaluating such integrals is that we apply the trapezoidal rule in the x-direction and then Simpson'1/3 or 3/8 rule in the y-direction. The following examples will illustrate the procedure very well.

Example

Evaluate the integral:

$$I = \int_{c}^{d} \int_{a}^{b} f(x,y)\,dx\,dy,$$ in which $a = 1.5$, $b = 3.0$ $c = 0.2$ and $d = 0.6$.

The numerical values for the function inside the integral are given in Table **19**.

Table 19. Multiple integrals.

y/x	0.1	0.2	0.3	0.4	0.5	0.6
0.5	0.165	0.428	0.678	0.942	1.190	1.431
1.0	0.271	0.640	1.003	1.359	1.703	2.035
1.5	0.447	0.990	1.524	2.045	2.549	3.031
2.0	0.738	1.568	2.384	3.177	3.943	4.672
2.5	1.216	2.520	3.800	5.044	6.241	7.379
3.0	2.005	4.090	6.136	8.122	10.030	11.841
3.5	3.306	6.679	9.986	13.196	16.277	19.198

Let us start with *x*-direction:

At $y = 0.2$

$$\int_{1.5}^{3.0} f(x,y)dx = \int_{1.5}^{3.0} f(x,0.2)dx = \frac{h}{2}(f_1 + 2\bullet[f_2 + f_3] + f_4)$$

$$\int_{1.5}^{3.0} f(x,y)dx = \frac{0.5}{2}(0.990 + 2\bullet[1.568 + 2.520] + 4.090) = 3.3140$$

At $y = 0.3$

$$\int_{1.5}^{3.0} f(x,y)dx = \int_{1.5}^{3.0} f(x,0.3)dx = \frac{h}{2}(f_1 + 2\bullet[f_2 + f_3] + f_4)$$

$$\int_{1.5}^{3.0} f(x,y)dx = \frac{0.5}{2}(1.524 + 2\bullet[2.384 + 3.800] + 6.136) = 5.007$$

Similarly, one can continue computations for the remaining values of y, leading to:

$y = 0.4$ $I = 6.6522$
$y = 0.5$ $I = 8.2368$
$y = 0.6$ $I = 9.7435$

Now take sum of these results in the y-direction according to Simpson's 1/3 rule:

$$I \cong \frac{h}{3}\left(f_{\substack{starting \\ value}} + f_{\substack{end \\ value}} + 2\bullet\left[\sum_{i=odd} f_i\right] + 4\bullet\left[\sum_{i=even} f_i\right] \right)$$

$$= \frac{0.1}{3}(3.314 + 9.7435 + 4\bullet[5.5.007 + 8.2368] + 2\bullet[6.6522])$$

$$= 2.6446$$

Example

Evaluate the integral $I = \int\limits_{4}^{4.6} \int\limits_{2}^{2.6} \left(\dfrac{1}{xy}\right) dxdy$, in which the step in x-direction $h = 0.2$ and the step in y-direction $k = 0.3$.

The numerical values for the function inside the integral are given in Table **20**.

Table 20. Data function.

x/y	4	4.2	4.4	4.6
2.0	0.125	0.119	0.114	0.109
2.3	0.109	0.104	0.099	0.095
2.6	0.096	0.092	0.087	0.084

Let us start with *x*-direction:

$y = 2.0$	$I = 0.0698$
$y = 2.3$	$I = 0.0608$
$y = 2.6$	$I = 0.0538$

Now take sum of these results in the y-direction according to Simpson's 1/3 rule:

$$I \cong \dfrac{k}{3}\left(f_{\substack{starting \\ value}} + f_{\substack{end \\ value}} + 2\bullet\left[\sum_{i=odd} f_i \right] + 4\bullet\left[\sum_{i=even} f_i \right]\right)$$

$$= 0.03667$$

17. INFINITE SERIES

A sequence u_n is a function defined on the set of natural numbers. This sequence is said to be ***convergent*** if there is a number ℓ with the property that for every $\varepsilon > 0$ there exists an integer N such that [39-41]:

$$|u_n - \ell| < \varepsilon \text{ for all } n > N.$$

We can write:

$$\lim_{n \to \infty} u_n = \ell$$

This sequence converges to ℓ.

Example

•The sequence 1, 4, 9, 16… is ***divergent*** because $\lim_{n \to \infty} n^2 = \infty$

The sequence $1, \dfrac{1}{e}, \dfrac{1}{e^2}, \dfrac{1}{e^3}, \ldots, \dfrac{1}{e^n}, \ldots$ is convergent because $\lim_{n \to \infty} \dfrac{1}{e^n} = 0$.

Consider the infinite series $u_n = u_1 + u_2 + u_3 + u_4 + \ldots$ and let their partial sums be $s_1, s_2, s_3 \ldots$

Where:

$s_1 = u_1$

$s_2 = u_1 = u_2$

$s_3 = u_1 + u_2 + u_3$

If $\lim_{n \to \infty} s_n = s$, we say that the sequence s_n converges to s.

Example

The Geometric Series

$$\sum_{n=1}^{\infty} \left(\frac{1}{2^n} \right) = \frac{1}{2} + \frac{1}{2^2} + \frac{1}{2^3} + \frac{1}{2^4} + \frac{1}{2^5} + \cdots$$

The sum of the first n terms:

$$S_n = \frac{1}{2} + \frac{1}{2^2} + \frac{1}{2^3} + \frac{1}{2^4} + \frac{1}{2^5} + \cdots + \frac{1}{2^n}$$

It can take the following form:

$$S_n = \left(1 - \frac{1}{2^n}\right)$$

The limit of the closed form of the nth terms equals one. Therefore, the geometric series is convergent.

17.1. Fundamentals of Infinite Series

(1) If u_n converges, then $\lim\limits_{n \to \infty} |u_n| = 0$. The converse, however, is not necessarily true.

(2) Multiplication of each term of a series by a constant different from zero does not affect the convergence or divergence.

(3) Removal (or addition) of finite number of terms from (or to) a series does not affect the convergence or divergence.

18. TESTS OF CONVERGENCE AND DIVERGENCE OF INFINITE SERIES

Before we use an infinite series for computational or other purposes, we must know whether it converges. In most cases that arise in engineering mathematics, this question may be answered by applying one of the following tests for convergence and divergence.

18.1. Comparison Test

• If $\sum u_n$ converges and $|v_n| \le |u_n|$ for all n, then $\sum u_n$ converges.

• If $\sum u_n$ diverges and $|v_n| \ge |u_n|$ for all n, then $\sum u_n$ converges

To simplify the problem, test, we introduce the following special series, see Table **21**:

Table 21. Some special series.

Series Name	P-series	Geometric Series
Series Form	$\sum\limits_{n=1}^{\infty} \dfrac{1}{n^p}$	$\sum\limits_{n=1}^{\infty} ar^{n-1}$
Divergent When	$p \le 1$	$\lvert r \rvert \ge 1$
Convergent	$p > 1$	$\lvert r \rvert < 1$

Example

Test the following series for convergence:

(a) $\sum\limits_{n=0}^{\infty} \dfrac{1}{2n+1}$

(b) $\sum\limits_{n=2}^{\infty} \dfrac{1}{\ln n}$

(c) $\sum\limits_{n=1}^{\infty} \dfrac{3^n}{n5^n}$

(d) $\sum\limits_{n=1}^{\infty} \dfrac{\ln n}{2n^n - 1}$

Solution

(a) $\sum\limits_{n=0}^{\infty} \dfrac{1}{2^n + 1}$

Since $\dfrac{1}{2^n+1} < \dfrac{1}{2^n}$

And

$\displaystyle\sum_{n=0}^{\infty}\dfrac{1}{2^n}$ Converges

Then $\displaystyle\sum_{n=0}^{\infty}\dfrac{1}{2^n+1}$ converges

(b) $\displaystyle\sum_{n=2}^{\infty}\dfrac{1}{\ln n}$

Since, $\dfrac{1}{\ln n} > \dfrac{1}{n}$ for all $n\geq 2$. And $\displaystyle\sum_{n=2}^{\infty}\dfrac{1}{n}$ diverges

Then $\displaystyle\sum_{n=2}^{\infty}\dfrac{1}{\ln n}$ also diverges.

18.2. Quotient Test

• If $\displaystyle\sum' u_n \geq 0$

And

$\displaystyle\sum' v_n \geq 0$

And if

$\displaystyle\lim_{n\to\infty}\dfrac{u_n}{v_n} = A \neq 0 \text{ or } \infty$

Then

$\left|u_n\right|$ and $\left|v_n\right|$ Either both converge or diverge

If $\lim\limits_{n\to\infty} \dfrac{u_n}{v_n} = A = 0$

And

$\sum' v_n$ Converges

Then

$\sum' u_n$ Converges

If $\lim\limits_{n\to\infty} \dfrac{u_n}{v_n} = A = \infty$

And

$\sum' v_n$ Diverges

Then

$\sum' u_n$ Diverges

Example

Test the following series for convergence:

(a) $\sum\limits_{n=1}^{\infty} \dfrac{n}{4n^3 - 2}$

(b) $\sum\limits_{n=2}^{\infty} \dfrac{\ln n}{\sqrt{n+1}}$

Solution

(a) $u_n = \sum\limits_{n=1}^{\infty} \dfrac{n}{4n^3 - 2}$

Choose, $v_n = \sum\limits_{n=1}^{\infty} \dfrac{1}{n^2}$

Applying the procedure of applying quotient test, we get:

$$\lim_{n \to 1} \frac{u_n}{v_n} = \lim_{n \to 1} \frac{\dfrac{n}{4n^3 - 2}}{\dfrac{1}{n^2}}$$

$$= \lim_{n \to 1} \left(\frac{n}{4n^3 - 2} \right)(n^2)$$

$$= \frac{1}{4}$$

18.3. Integral Test

The series $\sum\limits_{n=N}^{\infty} u_n$ in which $u_n = f(n)$ is positive, with continuous monotonic decreasing function, for all $n \geq N$, converge or diverge according to:

$$\int_{N}^{\infty} f(x)dx = \lim_{M \to \infty} \int_{N}^{M} f(x)dx$$

Exists or not respectively

Example

Test the following series for convergence:

(a) $\sum\limits_{n=1}^{\infty} \dfrac{2n}{n^3 - 1}$

(b) $\displaystyle\sum_{n=2}^{\infty} \frac{(\ln n)^2}{n}$

Solution

Let us apply the integral test, then:

(a) $I = \displaystyle\int_{2}^{\infty} \frac{2n}{n^2-1} \, dn = \left[\ln\left(n^2-1\right)\right]_{n=2}^{n=\infty} = \infty$

Then the series is divergent, since the integral is undefined.

The other part is left as an exercise.

Definition of Alternating Series

An alternating series is that series in which, the signs of terms alternate from positive to negative or vice versa.

18.4. Alternating Test

An alternating series is said to converge if the following two conditions are satisfied:

(1) $|u_{n+1}| \le |u_n|$

(2) $\displaystyle\lim_{n\to\infty} |u_n| = 0$

Example

Test the following series for convergence:

(a) $\displaystyle\sum_{n=1}^{\infty} \frac{(-1)^{n+1}}{n}$

(b) $\displaystyle\sum_{n=2}^{\infty} (-1)^{n-1} \frac{n}{n^2+1}$

Solution

(a) Since, the given series is alternative, we will apply the alternative series test:

$$u_n = \sum_{n=1}^{\infty} \frac{(-1)^{n+1}}{n}$$

$$\Rightarrow$$

$$|u_n| = \sum_{n=1}^{\infty} \frac{1}{n}$$

and

$$|u_{n+1}| = \sum_{n=1}^{\infty} \frac{1}{n+1}$$

Since, $\dfrac{1}{n+1} < \dfrac{1}{n}$ for all values of $n \geq 1$

And

$$\lim_{n \to \infty} |u_n| = \lim_{n \to \infty} \frac{1}{n} = 0.$$

Then

The series

$$u_n = \sum_{n=1}^{\infty} \frac{(-1)^{n+1}}{n} \quad \text{Converges}$$

Definition

The series $\sum' u_n$ is said to be absolutely convergent if $\sum |u_n|$ converges. If $\sum' u_n$ converges but $\sum |u_n|$ diverges, then the series $\sum' u_n$ is said to be conditionally convergent.

Example

Test the following series for convergence:

(a) $\displaystyle\sum_{n=1}^{\infty} \frac{(-1)^{n-1} n}{n^2 + 1}$

(b) $\displaystyle\sum_{n=2}^{\infty} (-1)^{n-1} \frac{1}{n \ln^2 n}$

Solution

Here:

$$u_n = \sum_{n=1}^{\infty} \frac{(-1)^{n-1} n}{n^2 + 1}$$

$$\Rightarrow$$

$$|u_n| = \sum_{n=1}^{\infty} \frac{n}{n^2 + 1}$$

$$\Rightarrow$$

$$|u_{n+1}| = \sum_{n=1}^{\infty} \frac{(n+1)}{(n+1)^2 + 1}$$

Since $|u_{n+1}| < |u_n|$ for all values of $n \geq 1$

And

$$\lim_{n \to \infty} |u_n| = 0.$$

Then, the series $\displaystyle\sum' u_n$ converges.

18.5. Ratio Test

Suppose that the following limit exists $\displaystyle\lim_{n \to \infty} \left| \frac{u_{n+1}}{u_n} \right| = L.$

Then the series $\displaystyle\sum' u_n$ precedes the following three cases:

•**Case (1)** Absolutely convergent if $L < 1$

•**Case (2)** Divergent if $L > 1$

•**Case (3)** If $L = 1$ the test fails

Example

Test the following series for convergence:

(a) $\displaystyle\sum_{n=1}^{\infty} n^4 e^{-n^2}$

(b) $\displaystyle\sum_{n=2}^{\infty} (-1)^{n-1} \frac{2^n}{n^2}$

Solution

Here:

$$u_n = \sum_{n=1}^{\infty} n^4 e^{-n^2} \Rightarrow |u_n| = \sum_{n=1}^{\infty} n^4 e^{-n^2}$$

and

$$|u_{n+1}| = \sum_{n=1}^{\infty} (n+1)^4 e^{-(n+1)^2}$$

$$\lim_{n\to\infty} \left| \frac{u_{n+1}}{u_n} \right| = \lim_{n\to\infty} \frac{(n+1)^4 e^{-(n+1)^2}}{(n)^4 e^{-(n)^2}} = 0 < 1$$

Therefore, the series is convergent.

19. TESTS OF SERIES OF FUNCTIONS

The above tests can be extended to the case where the un, are functions of x, dented by $\sum' u_n$.

In such case, the sequences or series will converge or diverge according to the particular values of x.

The set of values of x for which a sequence or series converges is called the region (domain) of convergence.

In this section, we are concerned with the power series as an important example for the series of functions.

19.1. Power Series

A series of the form:

$$a_o + a_1 x + a_2 x^2 + \ldots\ldots\ldots\ldots = \sum_{n=0}^{n=\infty} a_n x^n$$

Where $a_o, a_1, a_2, \ldots\ldots\ldots$ are constants, is called a ***power series*** in x. In general a power series converges for $|x| < R$, where the constant R is called the *radius of convergence* for the series. For $x = \pm R$ the series may or may not converge. There are two special cases $R = 0$ and $R \to \infty$ can arise.

Case (1)

$R = 0$ This case it converge only for $x = 0$

Case (2)

$R \to \infty$ This case it converge it converges for all x

Note that

(1) Similar remarks hold for a power series if x replaced by $(x - c)$ were c is constant.

(2) A power series can be differentiated or integrated term by term over an interval lying entirely within the interval of convergence.

(3) The sum of a convergent power series is continuous in any interval lying entirely within its interval of convergence.

(4) Two power series can be added or subtracted term for each value of x common to their interval of convergence.

(5) Two power series, for example $\sum\limits_{n=0}^{n=\infty} a_n x^n$ and $\sum\limits_{n=0}^{n=\infty} b_n x^n$ can be multiplied to obtain

a new series of the form: $\sum\limits_{n=0}^{n=\infty} c_n x^n$

(6) If the power series $\sum\limits_{n=0}^{n=\infty} a_n x^n$ is divided by the power series $\sum\limits_{n=0}^{n=\infty} b_n x^n$ where,

$b_o \neq 0$, the quotient can be written as a power series which converges for sufficiently small values of x.

Example

What value(s) for x makes the following series converge:

(a) $\sum\limits_{n=1}^{\infty} \dfrac{x^{n-1}}{n \times 3^n}$

(b) $\sum\limits_{n=1}^{\infty} (-1)^{n-1} \dfrac{x^{2n-1}}{(2n-1)!}$

Solution

Here

$$|u_n| = \sum\limits_{n=1}^{\infty} \dfrac{x^{n-1}}{n.3^n}$$

$\Rightarrow$

$$|u_{n+1}| = \sum\limits_{n=1}^{\infty} \dfrac{x^n}{(n+1).3^{(n+1)}}$$

To ensure absolute convergence of the series $\lim\limits_{n\to\infty} \left|\dfrac{u_{n+1}}{u_n}\right| = L < 1$

Therefore:

$$\lim_{n\to\infty}\left|\frac{u_{n+1}}{u_n}\right| = \frac{|x|}{3}$$

Consequently, the series converges absolutely, when $|x| < 3$ and diverges when $|x| > 3$. If $|x| = 3$, hence the test fails.

When the test fails, we proceed as follows:

For $x = 3$

The series becomes: $\displaystyle\sum_{n=1}^{\infty}\frac{1}{3n} = \frac{1}{3}\sum_{n=1}^{\infty}\frac{1}{n} \Rightarrow$ Diverge

For $x = -3$

The series becomes: $\displaystyle\sum_{n=1}^{\infty}\frac{(-1)^{n-1}}{3n} = \frac{1}{3}\sum_{n=1}^{\infty}\frac{(-1)^{n-1}}{n} \Rightarrow$ Converge

Then the interval of convergence is $-3 \le x < 3$.

The other part is left as an exercise.

20. SERIES EXPANSION OF FUNCTIONS

Taylor's expansion of a function $f(x)$ about a point $x = 0$ defined as:

$$f(x) = f(a) + \frac{(x-a)}{1!}f'(x) + \frac{(x-a)^2}{2!}f''(x) \cdot + \frac{(x-a)^{n-1}}{(n-1)!}f^{(n-1)}(x) + R_n$$

In this expansion, the remainder term defined as:

$$R_n = \frac{(x-x_o)^n}{(n)!}f^{(n)}(x)$$

In which, $x < x_o < a$

Notes

Taylor series for any function will converge in some interval if $\lim\limits_{n\to\infty} R_n = 0$ in this interval.

The form of Taylor series is unique, *e.g.*, for $h(x) = f(g(x))$ is the same as for $f(x)$ when replacing x by $g(x)$ in the expansion of $f(x)$.

21. TAYLOR'S EXPANSION FOR SOME BASIC FUNCTIONS

$$e^x = 1 + x + \frac{x^2}{2!} + \frac{x^3}{3!} + \frac{x^4}{4!} + \cdots\cdots\cdots\cdots\cdots -\infty < x < \infty$$

$$\sin x = x - \frac{x^3}{3!} + \frac{x^5}{5!} - \frac{x^7}{7!} + -\cdots\cdots\cdots\cdots -\infty < x < \infty$$

$$\cos x = 1 - \frac{x^2}{2!} + \frac{x^4}{4!} - \frac{x^6}{6!} \cdots\cdots\cdots\cdots\cdots -\infty < x < \infty$$

$$\ln(1+x) = x - \frac{x^2}{2!} + \frac{x^3}{3!} - \frac{x^4}{4!} + -\cdots\cdots\cdots\cdots -1 < x \le 1$$

$$\tan^{-1} x = x - \frac{x^3}{3!} + \frac{x^5}{5!} - \frac{x^7}{7!} + -\cdots\cdots\cdots\cdots -1 \le x \le 1$$

Example

Evaluate the following integral correct to three decimal places $I = \int\limits_{x=0}^{x=1} \frac{1-e^{-x^2}}{x^2}\, dx$

Solution

Since Taylor's expansion for e^u is known, and by replacing each u by $-x^2$, we get:

$$e^{-x^2} = 1 - x^2 + \frac{x^4}{2!} - \frac{x^6}{3!} + \frac{x^8}{4!} + \cdots\cdots$$

Then

$$\frac{1-e^{-x^2}}{x^2} = 1 - \frac{x^2}{2!} + \frac{x^4}{3!} - \frac{x^6}{4!} + \cdots$$

Since the series converges for all values of x, therefore, the integral can now be written in the following form:

$$\int_0^1 \frac{1-e^{-x^2}}{x^2}dx = \left(x - \frac{x^3}{2!\times 3} + \frac{x^5}{3!\times 3} - \frac{x^7}{4!\times 7} + \cdots \right)_{x=0}^{x=1}$$

$$= 0.862$$

22. THE BINOMIAL EXPANSION

The binomial expansion of a function $f(x)$ having the form $(1+x)^p$ can be written as:

$$(1+x)^p = 1 + px + \frac{p(p-1)}{2!}x^2 + \dots + \frac{p(p-1)\dots(p-n+1)}{n!}x^n + \dots$$

Where:

(a) If p is a positive integer or zero, the series terminates

(b) If p>0 but is not an integer, the series converges (absolutely) for $-1 \le x \le 1$.

(c) If $-1 < p < 0$, the series converges for $-1 < x \le 1$

(d) If $p \le -1$ the series converges for $-1 < x < 1$

Example

Prove that: $\ln(1+x) = x - \frac{x^2}{2} + \frac{x^3}{3} - \frac{x^4}{4} + \dots$

Solution

The binomial expansion for $\frac{1}{1+x}$ written as:

$$\frac{1}{1+x} = 1 - x + x^2 - x^3 + x^4 - + \dots$$

But

$$\int_0^x \frac{dx}{1+x} = \ln(1+x) = \int_0^x \left(1 - x + x^2 - x^3 + x^4 - + \dots\right) dx$$

$$\Rightarrow$$

$$\ln(1+x) = x - \frac{x^2}{2} + \frac{x^3}{3} - \frac{x^4}{4} + \dots$$

SUPPLEMENTARY PROBLEMS

Problem (1)

Test the convergence of the following series:

(1) $\displaystyle\sum_{n=1}^{\infty} \sqrt{\frac{n-1}{n+1}}$

(2) $\displaystyle\sum_{n=1}^{\infty} \frac{n^n}{n!}$

(3) $\displaystyle\sum_{n=0}^{\infty} \frac{1}{n!}$

(4) $1 + \dfrac{1}{\sqrt{2}} + \dfrac{1}{\sqrt{3}} + \dfrac{1}{\sqrt{4}} + \ldots\ldots$

(5) $\dfrac{1}{1.2} + \dfrac{1}{2.3} + \dfrac{1}{3.4} + \ldots\ldots$

(6) $\displaystyle\sum_{n=0}^{\infty} \left(\frac{4}{6}\right)^n$

(7) $\displaystyle\sum_{n=1}^{\infty} \frac{1}{n^2 + 1}$

(8) $\displaystyle\sum_{n=1}^{\infty} \frac{n}{4n^2 - 3}$

(9) $\displaystyle\sum_{n=1}^{\infty} \frac{n+2}{(n+1)\sqrt{(n-3)}}$

(10) $\displaystyle\sum_{n=1}^{\infty} \frac{n1}{n} \left(\frac{3}{7}\right)^n$

(11) $\displaystyle\sum_{n=1}^{\infty}\frac{1}{5n-3}$

(12) $\displaystyle\sum_{n=1}^{\infty}\frac{2n-1}{(3n+2)n^{4/3}}$

(13) $\displaystyle\sum_{n=1}^{\infty}\frac{n^2}{2n^3-1}$

(14) $\displaystyle\sum_{n=1}^{\infty}\sqrt{\frac{n-\ln n}{n^2-10n^3}}$

(15) $\displaystyle\sum_{n=1}^{\infty}\frac{4n^2+5n+9}{n(n^2+1)^{3/2}}$

(16) $\displaystyle\sum_{n=1}^{\infty}\frac{1}{n(\ln n)^3}$

(17) $\displaystyle\sum_{n=1}^{\infty}\frac{e^{-\tan^{-1}n}}{n^{2+1}}$

(18) $\displaystyle\sum_{n=1}^{\infty}\frac{1}{n\ln n}$

(19) $\displaystyle\sum_{n=1}^{\infty}\frac{3^n}{n^3}$

(20) $\displaystyle\sum_{n=1}^{\infty}\frac{10^{2n}}{(2n-1)!}$

Problem (2)

Test the following series for absolute or conditional convergence:

(1) $\displaystyle\sum_{n=1}^{\infty}\frac{2n-1}{(3n+2)n^{4/3}}$

(2) $\displaystyle\sum_{n=1}^{\infty} \frac{(-1)^n}{(n+1)e^n}$

(3) $\displaystyle\sum_{n=1}^{\infty} \frac{(-1)^n 4^n}{(n-1)!}$

(4) $\displaystyle\sum_{n=1}^{\infty} \frac{(-1)^n 2^n n1}{n^n}$

(5) $\displaystyle\sum_{n=1}^{\infty} \frac{(-1`)^n n^3}{(n^2+1)^{4/3}}$

(6) $\displaystyle\sum_{n=1}^{\infty} \frac{(-1)^{n-1}}{2^n-1}\sin\frac{1}{\sqrt{n}}$

Problem (3)

Determine the region of convergence for each of the following power series:

(1) $1 + \dfrac{x}{3.2} + \dfrac{x^2}{3^2.3} + \dfrac{x^3}{3^4.4} + \dots\dots$

(2) $1 - \dfrac{x}{5\sqrt{2}} + \dfrac{x^3}{5^2\sqrt{3}} - \dfrac{x^5}{5^3\sqrt{4}} + \dots\dots$

(3) $\dfrac{(2x-3)}{1} - \dfrac{(2x-3)^2}{3} + \dfrac{(2x-3)^3}{5} - \dots\dots$

(4) $\displaystyle\sum_{n=1}^{\infty} \frac{x^n}{n!}$

Problem (4)

Expand the following integrands into power series, then evaluate their corresponding integrals correct to four decimal places:

(1) $\displaystyle\int_{0}^{0.2}\frac{e^{x}-1}{x}dx$

(2) $\displaystyle\int_{0}^{0.1}\frac{\sin x}{x}dx$

Problem (5)

Expand each of the following function into power series. Mention the interval of convergence for each series.

(a) $\ln\left(\dfrac{1-x}{1+x}\right)$

(b) $\dfrac{1}{x+\sqrt{3}}$

(c) $\dfrac{x+2}{(x+4)(x-2)}$

Problem (6)

The following Table is for $f(x)=(1+\ln x)$. Determine estimates of its first derivative at $x = 0.15$, $x = 0.19$ and 0.23 using:

(a)   One term(b)Two terms

(c)   Three terms(d)Four terms

By comparing the analytical values, determine the errors of each estimate.

x	$(1+\ln x)$	Δ	Δ^2	$\Delta^3 x$
0.15	0.1761			
		0.0543		

Table contd….

0.17	0.2304		-0.0059	
		0.0484		0.0009
0.19	0.2788		-0.0050	
		0.0434		0.0011
0.21	0.3222		-0.0039	
		0.0395		0.0006
0.23	0.3617		-0.0033	
		0.0362		0.0005
0.25	0.3979		-0.0027	
		0.0335		0.0002
0.27	0.4314		-0.0025	
		0.0310		0.0005
0.29	0.4624		-0.0020	
		0.0290		
0.31	0.4914			

Problem (7)

The following Table is for $f(x) = \sin x$. Determine estimates of its first derivative at $x = 0.9$ using:

(b)   One term(b)Two terms

(d)   Three terms(d)Four terms

By comparing the analytical values, determine the errors of each estimate.

x	$\sin x$	x	$\sin x$
0.800	0.71736	0.901	0.78395
0.850	0.75128	0.902	0.78457
0.880	0.77074	0.905	0.78643
0.890	0.77707	0.910	0.78950
0.895	0.78021	0.920	0.79560
0.898	0.78208	0.950	0.81342
0.899	0.78270	1.000	0.84147

Problem (8)

Compute the first and the second derivative for $f(x) = \exp(x)$, using $h = 0.1$ at $x = 1.8$, and then compare the results with the exact values.

Problem (9)

Compute the first and the second derivative for $f(x) = \cos \pi x$, at $x = 0.45$ with the values of the given function at $x = 0, 0.25, 0.50, 0.75$ and 1.0, and then compare the results with the exact values.

Problem (10)

Construct the Lozenge's diagram for the three problems above, and then find:

   (a)   The first derivative

   (b)   The second derivative

Problem (11)

Evaluate the following integrals:

(1) $I = \int\limits_{0}^{\frac{\pi}{4}} \left(\dfrac{x}{\cos^2 x} \right) dx$

(2) $I = \int\limits_{0}^{\frac{\pi}{4}} \left(\dfrac{x}{\sin^2 x} \right) dx$

(3) $I = \int\limits_{0}^{\frac{\pi}{4}} \left(\dfrac{x}{\tan^2 x} \right) dx$

(a)   Trapezoidal rule

(b)   Simpson's 1/3 rule

Problem (12)

For the following tabulated function, find the integration of the function between the initial and the terminated points, given in Table.

x	$f(x)$	x	$f(x)$
0.0	93	0.6	35
0.1	87	0.7	39
0.2	68	0.8	48
0.3	55	0.9	53
0.4	42		
0.5	37		

Problem (13)

Use Romberg method to evaluate the following integrals up to three decimal places:

(a) $\quad I = \int\limits_{0}^{2} \sqrt{\tan x}\ dx$

(b) $\quad I = \int\limits_{1}^{2} \exp\!\left(x^2\right) dx$

Problem (14)

Evaluate the following integral:

$$I = \int\limits_{-0.2}^{0.6} \int\limits_{0.1}^{0.7} e^x \sin y\ dx$$

Problem (15)

Use Simpson'1/3 rule to evaluate the following integrals:

(a) $\quad I = \int\limits_{0}^{1} \frac{\sin x}{x}\ dx$

(b) $\quad I = \int\limits_{0}^{1} \frac{\cos x}{x}\ dx$

CHAPTER 3

Numerical Solution for System of Linear Equations

Abstract: The present chapter concerns mainly on major topics related to linear system of algebraic equations. These topics include the basic information about their constitution, Algebra, major types of numerical methods of solution, convergency and stability.

Keywords: Bisection, Determinants, False position, Gauss elimination, Gauss-Jordan elimination, Matrices.

1. INTRODUCTION

Solution of many applications lead to the system of linear algebraic equations, for example, the numerical solution for solving partial differential equations lead to a large system of algebraic linear equations [42-47].

There are what is called direct methods [48] and iterative methods [49] for solving the linear systems.

The present chapter starts with the basic informations, rules and different mathematical properties of the determinants. Determinants followed by matrices in some details and different types of matrices with different and brief examples.

After that different elementary direct and iterative methods are described in some details with illustrative examples.

2. DETERMINANTS

Major expressions occurring in different types of engineering systems can be modeled and solved by expressing them in the form of determinants.

A brief description of the important properties of determinants with applications to the solution of a system of linear equations is given in this chapter.

2.1. Definition

To give brief definition of the determinant, let us consider the following system of two linear equations [50]:

$$a_{11}x + a_{12}y = 0$$
$$a_{21}x + a_{22}y = 0$$

(1)

By Eliminating the variables x and y leads to:

$$a_{11}a_{22} - a_{21}a_{12} = 0$$

(2)

In a determinant form, equation (2) takes the following form:

$$\Delta = \begin{vmatrix} a_{11} & a_{12} \\ a_{21} & a_{22} \end{vmatrix}$$

(3)

Equation (3) is called a *determinant of second order*. In general a determinant of order n is denoted by:

$$\begin{vmatrix} a_{11} & a_{12} & a_{13} & a_{1n} \\ a_{21} & a_{22} & a_{23} & a_{2n} \\ \cdots\cdots\cdots\cdots\cdots\cdots\cdots\cdots\cdots \\ a_{n1} & a_{n2} & a_{n3} & a_{nn} \end{vmatrix}$$

(4)

In equation (4), the elements $a_{11}, a_{22}, a_{33}, ..., a_{nn}$ are called the *leading or principal diagonal.*

2.2. Minors

Each element in any determinant has a minor and is called the minor of an element [51]. The minor of an element is defined as the determinant obtained by deleting the row and the column which intersect in that element, and usually written in a capital letter.

Example

Find the minor of the element a_{32} in the following determinant:

$$\Delta = \begin{vmatrix} a_{11} & a_{12} & a_{13} \\ a_{21} & a_{22} & a_{23} \\ a_{31} & a_{32} & a_{33} \end{vmatrix}$$

Solution

The minor of the element a_{32} is obtained by deleting the third row and second column, therefore, the required minor will be:

$$A_{23} = \begin{vmatrix} a_1 & c_1 \\ a_2 & c_2 \end{vmatrix}$$

2.3. Cofactor

The cofactor of an element is defined as *the minor of that element with proper sign* $(-1)^{i+j}$, *where i is the i th row and j is the j th column of that element* [52].

Example

Find the cofactor of the element a_{32} in the following determinant:

$$\Delta = \begin{vmatrix} a_{11} & a_{12} & a_{13} \\ a_{21} & a_{22} & a_{23} \\ a_{31} & a_{32} & a_{33} \end{vmatrix}$$

Solution

First, the minor of the element a_{32} from previous example is:

$$A_{23} = \begin{vmatrix} a_1 & c_1 \\ a_2 & c_2 \end{vmatrix}$$

Then, the cofactor will take the following form:

$$A_{23} = (-1)^{2+3} \begin{vmatrix} a_1 & c_1 \\ a_2 & c_2 \end{vmatrix}$$

$$= (-1)^5 \begin{vmatrix} a_1 & c_1 \\ a_2 & c_2 \end{vmatrix}$$

$$= - \begin{vmatrix} a_1 & c_1 \\ a_2 & c_2 \end{vmatrix}$$

2.4. Laplace's Expansion

One of the methods to evaluate the value of any determinant is Laplace's expansion method [53], which simple can be stated as *"Multiply each element of the row (or column) in terms intend expanding the determinant, by its cofactor and then add up all these terms"*.

Example

Find the value of the following determinant:

$$\begin{vmatrix} 0 & 1 & 2 & 3 \\ 1 & 0 & 3 & 0 \\ 2 & 3 & 0 & 1 \\ 3 & 0 & 1 & 2 \end{vmatrix}$$

Solution

Choose the row or column with the maximum number of zeros. This is the case of the second row, therefore, expanding through that row, leads to:

$$\Delta = (-1)\begin{vmatrix} 1 & 2 & 3 \\ 3 & 0 & 1 \\ 0 & 1 & 2 \end{vmatrix} + 0 - 3\begin{vmatrix} 0 & 1 & 3 \\ 2 & 3 & 1 \\ 3 & 0 & 2 \end{vmatrix} + 0$$

Expanding the determinants, leads to:

$$\Delta = (-1)\left\{ \begin{vmatrix} 0 & 1 \\ 1 & 2 \end{vmatrix} - 2\begin{vmatrix} 3 & 1 \\ 0 & 2 \end{vmatrix} + 3\begin{vmatrix} 3 & 0 \\ 0 & 1 \end{vmatrix} \right\}$$

$$+ 3\left\{ 0\begin{vmatrix} 3 & 1 \\ 0 & 2 \end{vmatrix} - \begin{vmatrix} 2 & 1 \\ 3 & 2 \end{vmatrix} + 3\begin{vmatrix} 2 & 3 \\ 3 & 0 \end{vmatrix} \right\} = 88$$

3. PROPERTIES OF DETERMINANTS

(1) The value of a determinant remains the same if rows and columns are interchanged, *i.e*, $|A| = |A^T|$

(2) If all elements of any row (or column) are zeros except one element, then the value of the determinant is equal to the product of that element by its cofactor, for example:

$$|\Delta| = \begin{vmatrix} 0 & b_1 & 0 \\ a_2 & b_2 & c_2 \\ a_3 & b_3 & c_3 \end{vmatrix} = b_1 \begin{vmatrix} a_2 & c_2 \\ a_3 & c_3 \end{vmatrix}$$

(3) An interchange of any two rows (or columns) changes the sign of the determinant, for example:

$$\Delta = \begin{vmatrix} a_1 & b_1 & c_1 \\ a_2 & b_2 & c_2 \\ a_3 & b_3 & c_3 \end{vmatrix} = - \begin{vmatrix} a_2 & b_2 & c_2 \\ a_1 & b_1 & c_1 \\ a_3 & b_3 & c_3 \end{vmatrix}$$

And

$$\Delta = \begin{vmatrix} a_1 & b_1 & c_1 \\ a_2 & b_2 & c_2 \\ a_3 & b_3 & c_3 \end{vmatrix} = - \begin{vmatrix} b_1 & a_1 & c_1 \\ b_2 & a_2 & c_2 \\ b_3 & a_3 & c_3 \end{vmatrix}$$

(4) If all elements in any row (or column) are multiplied by a number, the determinant is also multiplied by this number, for example:

$$\Delta = \begin{vmatrix} a_1 & kb_1 & c_1 \\ a_2 & kb_2 & c_2 \\ a_3 & kb_3 & c_3 \end{vmatrix} = k \begin{vmatrix} a_1 & b_1 & c_1 \\ a_2 & b_2 & c_2 \\ a_3 & b_3 & c_3 \end{vmatrix}$$

(5) If any two rows (or columns) are the same or proportional the determinant is zero, for example:

$$\Delta = \begin{vmatrix} a_1 & a_1 & c_1 \\ a_2 & a_2 & c_2 \\ a_3 & a_3 & c_3 \end{vmatrix} = 0$$

(6) If we express the elements of each row (or column) as the sum of two of two terms, then the determinant can be expressed as the sum of two determinants having the same order, for example:

$$\Delta = \begin{vmatrix} a_1 + b_1 & c_1 & d_1 \\ a_2 + b_2 & c_2 & d_2 \\ a_3 + b_3 & c_3 & d_3 \end{vmatrix} = \begin{vmatrix} a_1 & c_1 & d_1 \\ a_2 & c_2 & d_2 \\ a_3 & c_3 & d_3 \end{vmatrix} + \begin{vmatrix} b_1 & c_1 & d_1 \\ b_2 & c_2 & d_2 \\ b_3 & c_3 & d_3 \end{vmatrix}$$

(7) The multiplication of elements of any row (or column) by a given number and added to corresponding elements of any other row (or column) does not change the value of the determinant, for example:

$$\begin{vmatrix} a_1 & b_1 & c_1 \\ a_2 & b_2 & c_2 \\ a_3 & b_3 & c_3 \end{vmatrix} = \begin{vmatrix} ka_1 + a_2 & kb_1 + b_2 & kc_1 + c_2 \\ a_2 & b_2 & c_2 \\ a_3 & b_3 & c_3 \end{vmatrix}$$

(8) If A and B are square matrices of the same order, then

$$|AB| = |A||B|$$

(9) The sum of the product of the elements of any row (or column) by the cofactors of another row (or column) is zero. If for a given determinant, for example:

$$\Delta = \begin{vmatrix} a_1 & b_1 & c_1 \\ a_2 & b_2 & c_2 \\ a_3 & b_3 & c_3 \end{vmatrix} = a_1 A_2 + b_1 B_2 + c_1 C_2 = 0$$

4. MATRICES

The matrix can be defined as a collocation of numbers or elements arranged in m rows and n columns and bounded by the bracket ( ) or [ ] is called $m \times n$ matrix.

The matrix is usually denoted by a capital letter, for example the matrix A of m rows and n columns is written as [54]:

$$A = \begin{pmatrix} a_{11} & a_{12} & a_{13} & a_{1j} & ..... & a_{1n} \\ a_{21} & a_{22} & a_{23} & a_{2j} & ..... & a_{2n} \\ .... \\ a_{m1} & a_{m2} & a_{m3} & a_{mj} & ..... & a_{mn} \end{pmatrix}$$

5. SPECIAL MATRICES

5.1. Row and Column Matrices

A matrix having a single row is called a **row matrix**, for example:

$$A_{1\times3} = \begin{pmatrix} 1 & 2 & 5 \end{pmatrix}$$

A matrix having a single column is called a **column matrix**, for example:

$$A_{3\times1} = \begin{pmatrix} 1 \\ 4 \\ 7 \end{pmatrix}$$

5.2. Square Matrix

The square matrix is characterized by the equal number of rows and columns, for example:

$$A_{4\times4} = \begin{pmatrix} 0 & 1 & 2 & 3 \\ 1 & 0 & 3 & 0 \\ 2 & 3 & 0 & 1 \\ 3 & 0 & 1 & 2 \end{pmatrix}$$

5.3. Diagonal Matrix

A square matrix all of whose elements except those in the leading diagonal are zero is called a diagonal matrix, for example:

$$A_{4\times4} = \begin{pmatrix} 2 & 0 & 0 & 0 \\ 0 & 5 & 0 & 0 \\ 0 & 0 & 7 & 0 \\ 0 & 0 & 0 & 3 \end{pmatrix}$$

5.4. Unit Matrix

A square matrix all of whose elements except those in the leading diagonal are zero is called a diagonal matrix such that the diagonal elements are unity, for example:

$$I_{4\times4} = \begin{pmatrix} 1 & 0 & 0 & 0 \\ 0 & 1 & 0 & 0 \\ 0 & 0 & 1 & 0 \\ 0 & 0 & 0 & 1 \end{pmatrix}$$

5.5. Null Matrix

If all elements of a given matrix are zero, it is called a null matrix and is denoted by O, for example:

$$O_{4\times4} = \begin{pmatrix} 0 & 0 & 0 & 0 \\ 0 & 0 & 0 & 0 \\ 0 & 0 & 0 & 0 \\ 0 & 0 & 0 & 0 \end{pmatrix}$$

5.6. Symmetric and Skew-Symmetric Matrices

A square matrix $A = a_{ij}$ is said to symmetric if $a_{ij} = a_{ji}$ for all i and j. If $a_{ij} = -a_{ji}$ for all i and j so that all the leading diagonal elements are zero, then the matrix is called a skew-symmetric matrix, for example, the matrix A below is symmetric the matrix B is skew-symmetric.

$$A = \begin{pmatrix} a & h & g \\ h & b & f \\ g & f & c \end{pmatrix}$$

$$B = \begin{pmatrix} 0 & h & -g \\ -h & 0 & f \\ g & -f & 0 \end{pmatrix}.$$

5.7. Triangular Matrices

The square matrix all of whose elements **below** the leading diagonal are zero is called an **upper triangular matrix**, for example:

$$A = \begin{pmatrix} a & h & g \\ 0 & b & f \\ 0 & 0 & c \end{pmatrix}$$

The square matrix all of whose elements **above** the leading diagonal are zero is called an **lower triangular matrix**, for example:

$$B = \begin{pmatrix} 1 & 0 & 0 \\ 2 & 3 & 0 \\ 1 & 0 & 4 \end{pmatrix}$$

6. ADDITION AND SUBTRACTION OF MATRICES

For a given two squares matrices A and B, the sum $A + B$ is the sum of the corresponding elements of A and B, for example:

$$A = \begin{pmatrix} a_1 & b_1 \\ a_2 & b_2 \\ a_3 & b_3 \end{pmatrix}$$

$$B = \begin{pmatrix} c_1 & d_1 \\ c_2 & d_2 \\ c_3 & d_3 \end{pmatrix}$$

$$A + B = \begin{pmatrix} a_1 + c_1 & b_1 + d_1 \\ a_2 + c_2 & b_2 + d_2 \\ a_3 + c_3 & b_3 + d_3 \end{pmatrix}$$

Similar procedure is carried out for subtraction of the same square matrices and is denoted by $A - B$ but replacing the $+$ sign by the opposite one $-$ve sign, for example:

$$A = \begin{pmatrix} a_1 & b_1 \\ a_2 & b_2 \\ a_3 & b_3 \end{pmatrix}$$

$$B = \begin{pmatrix} c_1 & d_1 \\ c_2 & d_2 \\ c_3 & d_3 \end{pmatrix}$$

$$A - B = \begin{pmatrix} a_1 - c_1 & b_1 - d_1 \\ a_2 - c_2 & b_2 - d_2 \\ a_3 - c_3 & b_3 - d_3 \end{pmatrix}$$

7. MULTIPLICATION OF A MATRIX BY A SCALAR

The product of a matrix A by a scalar k is a matrix produce a matrix whose, each element is k times the corresponding elements of A, for example:

$$A = \begin{pmatrix} a_1 & b_1 \\ a_2 & b_2 \\ a_3 & b_3 \end{pmatrix}$$

Then

$$kA = \begin{pmatrix} ka_1 & kb_1 \\ ka_2 & kb_2 \\ ka_3 & kb_3 \end{pmatrix}$$

8. MULTIPLICATION OF MATRICES

The multiplication of two matrices requires one and only one condition is that the number of columns in the first one is equal to the number of rows in the second, for example:

$$A_{4\times4} = \begin{pmatrix} a_1 & b_1 & c_1 \\ a_2 & b_2 & c_2 \\ a_3 & b_3 & c_3 \\ a_4 & b_4 & c_4 \end{pmatrix}$$

$$B_{3\times2} = \begin{pmatrix} l_1 & l_2 \\ m_1 & m_2 \\ n_1 & n_2 \end{pmatrix}$$

Then the resultant matrix will be:

$$A_{4\times4} \times B_{3\times2} = \begin{pmatrix} a_1 l_1 + b_1 m_1 + c_1 n_1 & a_1 l_2 + b_1 m_2 + c_1 n_2 \\ a_2 l_1 + b_2 m_1 + c_2 n_1 & a_2 l_2 + b_2 m_2 + c_2 n_2 \\ a_3 l_1 + b_3 m_1 + c_3 n_1 & a_3 l_2 + b_3 m_2 + c_3 n_2 \\ a_4 l_1 + b_4 m_1 + c_4 n_1 & a_4 l_2 + b_4 m_2 + c_4 n_2 \end{pmatrix}_{4\times2}$$

9. TRANSPOSE OF A MATRIX

By interchanging rows and columns in a given matrix, the resultant matrix is called the transpose of A and is denoted by A^T [55].

10. ADJOINT OF A SQUARE MATRIX

Two major steps to find the adjoint of a given square matrix, the first step is to form the cofactor matrix then the transpose of the cofactor matrix is the adjoint matrix.

Example

Find the adjoint matrix for the given matrix A:

$$A = \begin{pmatrix} a_1 & b_1 & c_1 \\ a_2 & b_2 & c_2 \\ a_3 & b_3 & c_3 \end{pmatrix}$$

Solution

Following up the procedure above and let us start by forming the cofactors matrix as follow:

$$C.M. = \begin{pmatrix} A_1 & B_1 & C_1 \\ A_2 & B_2 & C_2 \\ A_3 & B_3 & C_3 \end{pmatrix}$$

The transpose of cofactor matrix is itself the required adjoint matrix is given by:

$$adj(A) = \left(C.M.\right)^T = \begin{pmatrix} A_1 & A_2 & A_3 \\ B_1 & B_2 & B_3 \\ C_1 & C_2 & C_3 \end{pmatrix}$$

Example

Find the adjoint matrix for:

$$A = \begin{pmatrix} 1 & 4 & 5 \\ 3 & 6 & 9 \\ 7 & 1 & 2 \end{pmatrix}$$

Solution

(1) Form the cofactor matrix

$$C.M. = \begin{pmatrix} +\begin{vmatrix} 6 & 9 \\ 1 & 2 \end{vmatrix} & -\begin{vmatrix} 3 & 9 \\ 7 & 2 \end{vmatrix} & +\begin{vmatrix} 3 & 6 \\ 7 & 1 \end{vmatrix} \\ -\begin{vmatrix} 4 & 5 \\ 1 & 2 \end{vmatrix} & +\begin{vmatrix} 1 & 5 \\ 7 & 2 \end{vmatrix} & -\begin{vmatrix} 1 & 4 \\ 7 & 1 \end{vmatrix} \\ +\begin{vmatrix} 4 & 5 \\ 6 & 9 \end{vmatrix} & -\begin{vmatrix} 1 & 5 \\ 3 & 9 \end{vmatrix} & +\begin{vmatrix} 1 & 4 \\ 3 & 6 \end{vmatrix} \end{pmatrix}$$

$$= \begin{pmatrix} 3 & 57 & -39 \\ -3 & -33 & 28 \\ 6 & 6 & -6 \end{pmatrix}$$

Then the adjoint matrix is

$$adj(A) = \begin{pmatrix} 3 & -3 & 6 \\ 57 & -33 & 6 \\ -39 & 28 & -6 \end{pmatrix}$$

11. INVERSE OF A MATRIX

If the matrix A be any a square matrix, then the matrix B, if it exists under the condition $AB = BA = I$, then the matrix B is called the inverse of A and take the notation A^{-1}, I being a unit matrix [56].

Example

Find the inverse of $\begin{pmatrix} 1 & 1 & 3 \\ 1 & 3 & -3 \\ -2 & -4 & -4 \end{pmatrix}$

Solution

The determinant of the given matrix is:

$$\Delta = \begin{vmatrix} 1 & 1 & 3 \\ 1 & 3 & -3 \\ -2 & -4 & -4 \end{vmatrix} = \begin{vmatrix} a_1 & b_1 & c_1 \\ a_2 & b_2 & c_2 \\ a_3 & b_3 & c_3 \end{vmatrix}$$

Following up the procedure to get cofactors, leads to:

$A_1 = 24 \quad\quad A_2 = -8 \quad\quad A_3 = -12$

$B_1 = 10 \quad\quad B_2 = 2 \quad\quad B_3 = 6$

$C_1 = 2 \quad\quad C_2 = 2 \quad\quad C_3 = 2$

$$\Delta = a_1 A_1 + a_2 A_2 + a_3 A_3 = -8$$

$$adj(A) = \begin{pmatrix} A_1 & A_2 & A_3 \\ B_1 & B_2 & B_3 \\ C_1 & C_2 & C_3 \end{pmatrix} = \begin{pmatrix} -24 & -8 & -12 \\ 10 & 2 & 6 \\ 2 & 2 & 2 \end{pmatrix}$$

Hence, the inverse of the given matrix will be:

$$\frac{adj(A)}{\Delta} = \frac{1}{-8} \begin{pmatrix} -24 & -8 & -12 \\ 10 & 2 & 6 \\ 2 & 2 & 2 \end{pmatrix}$$

12. SYSTEM OF LINEAR EQUATIONS

A linear system of algebraic equations can be represented in three different forms [57].

Assume that we have n system of linear algebraic equation then this system can be represented in one of the following forms:

12.1. Normal Form

$$a_{11}x_1 + a_{12}x_2 + \ldots\ldots\ldots + a_{1n}x_n = b_1$$
$$a_{21}x_1 + a_{22}x_2 + \ldots\ldots\ldots + a_{2n}x_n = b_2$$
$$\ldots\ldots\ldots\ldots\ldots\ldots\ldots\ldots\ldots\ldots\ldots\ldots\ldots$$
$$a_{n1}x_1 + a_{n2}x_2 + \ldots\ldots\ldots + a_{nn}x_n = b_n$$

12.2. Matrix Form

$$\begin{pmatrix} a_{11} & a_{12} & a_{13}............................a_{1n} \\ a_{21} & a_{22} & a_{23}............................a_{2n} \\ .... \\ a_{n1} & a_{n2} & a_{n3}............................a_{nn} \end{pmatrix} \begin{pmatrix} x_1 \\ x_2 \\ ... \\ ... \\ x_n \end{pmatrix} = \begin{pmatrix} b_1 \\ b_2 \\ ... \\ ... \\ b_n \end{pmatrix}$$

12.3. Compact Matrix Form

$Ax = b$

Where

A: Coefficient Matrix

x: Unknown Matrix

b: Absolute terms Matrix

13. CONSISTENCY OF A SYSTEM OF NON-HOMOGENEOUS LINEAR EQUATIONS

The system of non-homogeneous linear equations is consistent if and only if the coefficient matrix A and the augmented matrix K are of the same rank otherwise the system is inconsistent [58].

Given a system of m linear equations:

$$\left. \begin{array}{l} a_{11}x_1 + a_{12}x_2 + ... + a_{1n}x_n = k_1 \\ a_{21}x_1 + a_{22}x_2 + ... + a_{2n}x_n = k_2 \\ .... \\ a_{m1}x_1 + a_{m2}x_2 + ... + a_{mn}x_n = k_m \end{array} \right\}$$

$$[A]\underline{X} = \underline{K}$$

Where:

$[A]$ Coefficients matrix $m \times n$,

$\underline{X}$ A vector of order n,

$\underline{K}$ A vector of order m,

The above system is said to be consistent if and only if the determinants $[A]$ and $[A:K]$ are of the same rank. Consider the following three cases:

1. Rank of $[A]$ equals the rank of $[A:K]$ and $r = n$. In this case the vector $\underline{X}$ contains a unique solution.

2. Rank of A equals the rank of $\underline{K}$ and $r < n$. In this case

$$\underline{X} = \left[\frac{\underline{X}_r}{\underline{X}_{n-r}} \right]$$

Where

$$\underline{X}_r = \begin{bmatrix} x_1 \\ x_2 \\ \cdots \\ x_r \end{bmatrix}$$

$$\underline{X}_{n-r} = \begin{bmatrix} x_{r+1} \\ x_{r+2} \\ \cdots \\ x_n \end{bmatrix}$$

The $\underline{X}_{n-r}$ elements are arbitrarily chosen and $\underline{X}_r$ are functions of the $\underline{X}_{n-r}$ elements. In this case we have an infinite number of solutions.

Example

Test for consistency and solve the following system of equations:

$5x + 3y + 7z = 4$

$3x + 26y + 2z = 9$

$7x + 2y + 10z = 5$

Solution

We have

$$\begin{bmatrix} 5 & 3 & 7 \\ 3 & 26 & 2 \\ 7 & 2 & 10 \end{bmatrix} \begin{bmatrix} x \\ y \\ z \end{bmatrix} = \begin{bmatrix} 4 \\ 9 \\ 5 \end{bmatrix}$$

1. Operate $3R_1$, $5R_2$

$$\begin{bmatrix} 15 & 9 & 21 \\ 15 & 130 & 10 \\ 7 & 2 & 10 \end{bmatrix} \begin{bmatrix} x \\ y \\ z \end{bmatrix} = \begin{bmatrix} 12 \\ 45 \\ 5 \end{bmatrix}$$

2. Operate $R_2 - R_1$

$$\begin{bmatrix} 15 & 9 & 21 \\ 0 & 121 & -11 \\ 7 & 2 & 10 \end{bmatrix} \begin{bmatrix} x \\ y \\ z \end{bmatrix} = \begin{bmatrix} 12 \\ 33 \\ 5 \end{bmatrix}$$

3. Operate $7/8\ R_1$, $5R_3$, $1/11\ R_2$

$$\begin{bmatrix} 35 & 21 & 49 \\ 0 & 11 & -1 \\ 35 & 10 & 50 \end{bmatrix} \begin{bmatrix} x \\ y \\ z \end{bmatrix} = \begin{bmatrix} 28 \\ 3 \\ 25 \end{bmatrix}$$

4. Operate $R_3 - R_1 + R_2$, $3/7\ R_1$

$$\begin{bmatrix} 5 & 3 & 7 \\ 0 & 11 & -1 \\ 0 & 0 & 0 \end{bmatrix} \begin{bmatrix} x \\ y \\ z \end{bmatrix} = \begin{bmatrix} 4 \\ 3 \\ 0 \end{bmatrix}$$

The ranks of coefficient matrix and augmented matrix for the last set of equations are both two. Hence the equations are consistent. In addition, the given system is equivalent to:

$$5x + 3y + 7z = 4$$

$$11y - z = 3$$

Then

$$y = \frac{3}{11} + \frac{z}{11}$$

And

$$x = \frac{7}{11} - \frac{16}{11}z$$

Where z is a parameter

Here

$$x = \frac{7}{11} \ \& \ y = \frac{3}{11} \ \& \ z = 0$$

Is a particular solution

3. Rank of $[A]$ is not equal to the rank of $\underline{K}$ in this case there is no any possible solution.

Example

Investigate the values of λ and μ so that the equations:

$$2x + 3y + 5z = 9$$

$$7x + 3y - 2z = 8$$

$$2x + 3y + \lambda z = \mu$$

Have:

(i) No solution

(ii) A Unique solution

(iii) An infinite number of solutions.

Solution

We have:

$$\begin{bmatrix} 2 & 3 & 5 \\ 7 & 3 & -2 \\ 2 & 3 & \lambda \end{bmatrix} \begin{bmatrix} x \\ y \\ z \end{bmatrix} = \begin{bmatrix} 9 \\ 8 \\ \mu \end{bmatrix}$$

The system admits of a unique solution if and only if, $|A| \neq 0$

$$|A| = \begin{vmatrix} 2 & 3 & 5 \\ 7 & 3 & -2 \\ 20 & 3 & \lambda \end{vmatrix} = 2(3\lambda + 6) - 3(7\lambda + 40) + 5(21 - 6)$$

Case (1)

For a unique solution $\lambda \neq 5$ and μ may have any value.

Case (2)

If $\lambda = 5$. In this case the third row of the matrix A is equal to the first row. Hence the rank of A is two and the rank of K is three. Hence, the system will have no solution.

Case (3)

If $\lambda = 5$ and $\mu = 0$, the system will have an infinite number of solutions.

14. SYSTEM OF LINEAR HOMOGENEOUS EQUATIONS

A system of linear equations is said to be homogeneous if the right hand vector of the system $[A]\underline{X} = \underline{K}$ is zero. Consider the homogeneous linear equations.

$$a_{11}x_1 + a_{12}x_2 + \ldots\ldots + a_{1n}x_n = 0$$
$$a_{21}x_1 + a_{22}x_2 + \ldots\ldots + a_{2n}x_n = 0$$
$$\ldots\ldots\ldots\ldots\ldots\ldots\ldots\ldots\ldots\ldots\ldots\ldots\ldots\ldots\ldots$$
$$a_{n1}x_1 + a_{n2}x_2 + \ldots\ldots + a_{nn}x_n = 0$$

$$[\mathrm{A}]\underline{X} = 0$$

There are two possible cases:

Case (1)

If $|A| = 0$ then, $\underline{X}$ have an infinite number of solutions.

Case (2)

If $|A| \neq 0$ then, $\underline{X} = [0]$

15. CHARACTERISTIC EQUATION

If A is any square matrix of order n, we can form the matrix $A - \lambda I$, where λ is a scalar and I is the nth order unit matrix, and the determinant of this matrix equated zero [59].

$$|A - \lambda I| = \begin{vmatrix} a_{11} - \lambda & a_{12} & \ldots & a_{1n} \\ a_{21} & a_{22} - \lambda & \ldots & a_{2n} \\ \ldots & \ldots & \ldots & \ldots \\ a_{n1} & a_{n2} & \ldots & a_{nn} - \lambda \end{vmatrix} = 0$$

The expansion result is called the ***characteristic equation*** of A. Expanding the determinant, the characteristic equation takes the form:

$$(-1)^n \lambda^n + k_1 \lambda^{n-1} + k_2 \lambda^{n-2} + \ldots\ldots + k_n = 0$$

The roots of this equation are called the ***characteristic roots*** or ***eigenvalues*** of the matrix A.

16.1. EigenVectors

If $X_{r\times 1}$ is a column vector and $A_{n\times n}$ is a square matrix as follows [60]:

$$X = \begin{vmatrix} x_1 \\ x_2 \\ ... \\ x_r \end{vmatrix} \qquad \& \qquad A = \begin{vmatrix} a_{11}-\lambda & a_{12} & ... & a_{1n} \\ ... & ... & ... & ... \\ a_{n1} & a_{n2} & ... & a_{nn} \end{vmatrix}$$

Then the linear transformation $\underline{Y} = [A]\underline{X}$ carries the column vector X into the column vector Y by means of the square matrix $A_{n\times n}$. In practice, it is often required to find such vectors which, transform into themselves or to a scalar multiple of themselves. Let $X_{r\times 1}$ a vector which transforms into λX by means of the transformation $\underline{Y} = [A]\underline{X}$.

Then

$$\lambda X = AX$$
$$AX - \lambda IX = 0$$
$$[A - \lambda I]X = 0$$

This matrix equation represents n homogeneous linear equations of the form:

$$\left.\begin{array}{l} (a_{11}-\lambda)x1 + a_{12}x_2 + ... + a_{1n}x_n = 0 \\ a_{21}x_1 + (a_{22}-\lambda)x_2 + ... + a_{2n}x_n = 0 \\ \quad .... \qquad\qquad .... \qquad\qquad ... \qquad .. \\ a_{n1}x_1 + a_{n2}x_2 + ... + (a_{nn}-\lambda)x_n = 0 \end{array}\right\}$$

This system will have a non-trivial solution only if the coefficient matrix is singular, $A - \lambda I = 0$. This is called the characteristic equation of the ***transformation.***

<u>Example</u>

Find the characteristic equation of the matrix A given by:

$$A = \begin{vmatrix} 1 & 1 & 3 \\ 1 & 3 & -3 \\ -2 & -4 & -4 \end{vmatrix}$$

Solution

The characteristic equation is:

$$|A - \lambda I| = \begin{vmatrix} 1-\lambda & 1 & 3 \\ 1 & 3-\lambda & -3 \\ -2 & -4 & -4-\lambda \end{vmatrix} = 0$$

Then

$$\lambda^3 - 20\lambda + 8 = 0$$

This is the required equation.

Example

Find the eigenvalues and eigenvectors of the matrix.

$$A = \begin{vmatrix} 1 & 1 & 0 \\ 0 & 1 & 2 \\ 1 & 0 & -2 \end{vmatrix}$$

Solution

The characteristic equation is

$$|A - \lambda I| = \begin{vmatrix} 1-\lambda & 1 & 0 \\ 0 & 1-\lambda & 2 \\ 1 & 0 & -2-\lambda \end{vmatrix}$$

$$|A - \lambda I| = -\lambda^3 + 3\lambda = 0$$

$$\lambda\left(3-\lambda^2\right)=0$$

$$\lambda_1 = 0$$

$$\lambda_2 = \sqrt{3}$$

$$\lambda_3 = -\sqrt{3}$$

The corresponding eigenvectors are obtained through the following procedure:

$$(A-\lambda_i I)U^i = \begin{vmatrix} 1-\lambda_i & 1 & 0 \\ 0 & 1-\lambda_i & 2 \\ 1 & 0 & -2-\lambda_i \end{vmatrix} \begin{bmatrix} u^i_1 \\ u^i_2 \\ u^i_3 \end{bmatrix} = 0$$

(1)For $\lambda_1 = 0$

$$\begin{vmatrix} 1 & 1 & 0 \\ 0 & 1 & 2 \\ 1 & 0 & -2 \end{vmatrix} \begin{bmatrix} u^1_1 \\ u^1_2 \\ u^1_3 \end{bmatrix} = \begin{bmatrix} 0 \\ 0 \\ 0 \end{bmatrix}$$

Operate $R_3 - R_1$

$$\begin{vmatrix} 1 & 1 & 0 \\ 0 & 1 & 2 \\ 0 & 0 & 0 \end{vmatrix} \begin{bmatrix} u^1_1 \\ u^1_2 \\ u^1_3 \end{bmatrix} = \begin{bmatrix} 0 \\ 0 \\ 0 \end{bmatrix}$$

Then

$$\begin{bmatrix} 1 & 0 \\ 0 & 1 \end{bmatrix} \begin{bmatrix} u^1_1 \\ u^1_2 \end{bmatrix} = \begin{bmatrix} 2 \\ -2 \end{bmatrix}$$

Now take $u^1_3 = 1$ imply that:

$$u^1_1 = 2$$

And

$$u^1{}_2 = -2$$

Then the eigenvector corresponding to $\lambda_1 = 0$ can be written in a matrix form as follow:

$$U^1 = \begin{vmatrix} u^1{}_1 \\ u^1{}_2 \\ u^1{}_3 \end{vmatrix} = \begin{bmatrix} 2 \\ -2 \\ 1 \end{bmatrix}$$

(2)For $\lambda_2 = \sqrt{3}$

$$\begin{vmatrix} 1-\sqrt{3} & 1 & 0 \\ 0 & 1-\sqrt{3} & 2 \\ 1 & 0 & -2-\sqrt{3} \end{vmatrix} \begin{bmatrix} u^2{}_1 \\ u^2{}_2 \\ u^2{}_3 \end{bmatrix} = \begin{bmatrix} 0 \\ 0 \\ 0 \end{bmatrix}$$

Then

$$u^2{}_1 = \left(2+\sqrt{3}\right)u^2{}_3$$
$$u^2{}_2 = \left(1+\sqrt{3}\right)u^2{}_3$$

Now take $u^2{}_3 = 1$ imply that:

$$u^2{}_1 = 2+\sqrt{3}$$

And

$$u^2{}_2 = 1-\sqrt{3}$$

Then the eigenvector corresponding to $\lambda_2 = \sqrt{3}$ can be written in a matrix form as follow:

$$U^2 = \begin{bmatrix} u^2_1 \\ u^2_2 \\ u^2_3 \end{bmatrix} = \begin{bmatrix} 2+\sqrt{3} \\ 1+\sqrt{3} \\ 1 \end{bmatrix}$$

(3)For $\lambda_2 = -\sqrt{3}$

$$\begin{vmatrix} 1+\sqrt{3} & 1 & 0 \\ 0 & 1+\sqrt{3} & 2 \\ 1 & 0 & -2+\sqrt{3} \end{vmatrix} \begin{bmatrix} u^3_1 \\ u^3_2 \\ u^3_3 \end{bmatrix} = \begin{bmatrix} 0 \\ 0 \\ 0 \end{bmatrix}$$

Then

$$u^3_1 = \left(2-\sqrt{3}\right)u^3_3$$
$$u^3_2 = \left(-1+\sqrt{3}\right)u^3_3$$

Now take $u^3_3 = 1$ imply that:

$$u^3_1 = 2-\sqrt{3}$$

And

$$u^3_2 = -1+\sqrt{3}$$

Then the eigenvector corresponding to $\lambda_3 = -\sqrt{3}$ can be written in a matrix form as follow:

$$U^3 = \begin{bmatrix} u^3_1 \\ u^3_2 \\ u^3_3 \end{bmatrix} = \begin{bmatrix} 2-\sqrt{3} \\ 1-\sqrt{3} \\ 1 \end{bmatrix}$$

Example

Find the eigenvalues and eigen vectors of the matrix.

$$A = \begin{vmatrix} 8 & -6 & 2 \\ -6 & 7 & -4 \\ 2 & -4 & 3 \end{vmatrix}$$

Solution

The characteristic equation is

$$|A - \lambda I| = \begin{vmatrix} 8-\lambda & -6 & 2 \\ -6 & 7-\lambda & -4 \\ 2 & -4 & -2-\lambda \end{vmatrix} = 0$$

$$|A - \lambda I| = \lambda^3 + 18\lambda^2 - 45\lambda = 0$$

$$\lambda_1 = 0$$
$$\lambda_2 = 3$$
$$\lambda_3 = 15$$

$$(A - \lambda_i I)U^i = \begin{vmatrix} 8-\lambda_i & -6 & 2 \\ -6 & 7-\lambda_i & -4 \\ 2 & -4 & 3-\lambda_i \end{vmatrix} \begin{bmatrix} u^i_1 \\ u^i_2 \\ u^i_3 \end{bmatrix} = 0$$

(1)   For $\lambda_1 = 0$

$$\begin{vmatrix} 8 & -6 & 2 \\ -6 & 7 & -4 \\ 2 & -4 & 3 \end{vmatrix} \begin{bmatrix} u^1_1 \\ u^1_2 \\ u^1_3 \end{bmatrix} = \begin{bmatrix} 0 \\ 0 \\ 0 \end{bmatrix}$$

Operate $R_3 + R_2$

$$\begin{vmatrix} 8 & -6 & 2 \\ -4 & 3 & -1 \\ 2 & -4 & 3 \end{vmatrix} \begin{bmatrix} u^1_1 \\ u^1_2 \\ u^1_3 \end{bmatrix} = \begin{bmatrix} 0 \\ 0 \\ 0 \end{bmatrix}$$

Operate $R_1 + 2R_2$

$$\begin{vmatrix} 0 & 0 & 0 \\ -4 & 3 & -1 \\ 2 & -4 & 3 \end{vmatrix} \begin{bmatrix} u^1{}_1 \\ u^1{}_2 \\ u^1{}_3 \end{bmatrix} = \begin{bmatrix} 0 \\ 0 \\ 0 \end{bmatrix}$$

Operate $R_2 + 2R_3$

$$\begin{vmatrix} 0 & 0 & 0 \\ 0 & -5 & 5 \\ 2 & 1 & -2 \end{vmatrix} \begin{bmatrix} u^1{}_1 \\ u^1{}_2 \\ u^1{}_3 \end{bmatrix} = \begin{bmatrix} 0 \\ 0 \\ 0 \end{bmatrix}$$

Take 5 as a factor from the second row, gives:

$$\begin{vmatrix} 0 & 0 & 0 \\ 0 & -1 & 1 \\ 2 & 1 & -2 \end{vmatrix} \begin{bmatrix} u^1{}_1 \\ u^1{}_2 \\ u^1{}_3 \end{bmatrix} = \begin{bmatrix} 0 \\ 0 \\ 0 \end{bmatrix}$$

Operate $R_3 + R_2$

$$\begin{vmatrix} 0 & 0 & 0 \\ 0 & -1 & 1 \\ 2 & 0 & -1 \end{vmatrix} \begin{bmatrix} u^1{}_1 \\ u^1{}_2 \\ u^1{}_3 \end{bmatrix} = \begin{bmatrix} 0 \\ 0 \\ 0 \end{bmatrix}$$

Interchange the first and the third rows

$$\begin{vmatrix} 2 & 0 & -1 \\ 0 & -1 & 1 \\ 0 & 0 & 0 \end{vmatrix} \begin{bmatrix} u^1{}_1 \\ u^1{}_2 \\ u^1{}_3 \end{bmatrix} = \begin{bmatrix} 0 \\ 0 \\ 0 \end{bmatrix}$$

Then

$$u^1{}_1 = \frac{1}{2} u^1{}_3$$

$u^1{}_2 = u^1{}_3$

Now take $u^1{}_3 = 1$ imply that:

$u^1{}_1 = \dfrac{1}{2}$

And

$u^1{}_2 = 1$

Then

$$U^1 = \begin{bmatrix} u^1{}_1 \\ u^1{}_2 \\ u^1{}_3 \end{bmatrix} = \begin{vmatrix} \dfrac{1}{2} \\ 1 \\ 1 \end{vmatrix}$$

Similarly, for the second and the third eigenvalues, the corresponding eigenvectors are as follows:

$$U^2 = \begin{vmatrix} u^2{}_1 \\ u^2{}_2 \\ u^2{}_3 \end{vmatrix} = \begin{vmatrix} -1 \\ -\dfrac{1}{2} \\ 1 \end{vmatrix}$$

$$U^3 = \begin{vmatrix} u^3{}_1 \\ u^3{}_2 \\ u^3{}_3 \end{vmatrix} = \begin{bmatrix} 2 \\ -2 \\ 1 \end{bmatrix}$$

16. METHODS OF SOLUTION OF SYSTEM OF LINEAR EQUATIONS

16.1. Cramer's Rule

This method is quite simple and is used to solve a linear system of linear equations. This method requires unique condition is that the number of equations will be the same as the same as the number of unknowns [61].

Example

$$a_1 x + b_1 y + c_1 z = d_1$$

Given: $a_2 x + b_2 y + c_2 z = d_2$

$$a_3 x + b_3 y + c_3 z = d_3$$

Firstly; evaluate the value of the determinant of coefficients to ensure that its value will be a non zero,

$$\Delta = \begin{vmatrix} a_1 & b_1 & c_1 \\ a_2 & b_2 & c_2 \\ a_3 & b_3 & c_3 \end{vmatrix}$$

Secondly; for each unknown, obtain a determinant by replacing the corresponding column in the determinant of coefficients by the vector of the absolute terms,

$$\Delta x = \begin{vmatrix} d_1 & b_1 & c_1 \\ d_2 & b_2 & c_2 \\ d_3 & b_3 & c_3 \end{vmatrix}$$

Similarly,

$$\Delta y = \begin{vmatrix} a_1 & d_1 & c_1 \\ a_2 & d_2 & c_2 \\ a_3 & d_3 & c_3 \end{vmatrix}$$

And finally;

$$\Delta z = \begin{vmatrix} d_1 & b_1 & d_1 \\ d_2 & b_2 & d_2 \\ d_3 & b_3 & d_3 \end{vmatrix}$$

Thirdly; the values of the unknowns are obtained as follow:

$$x = \frac{\Delta x}{\Delta} = \frac{\begin{vmatrix} d_1 & b_1 & c_1 \\ d_2 & b_2 & c_2 \\ d_3 & b_3 & c_3 \end{vmatrix}}{\begin{vmatrix} a_1 & b_1 & c_1 \\ a_2 & b_2 & c_2 \\ a_3 & b_3 & c_3 \end{vmatrix}}$$

Similarly,

$$y = \frac{\Delta y}{\Delta} = \frac{\begin{vmatrix} a_1 & d_1 & c_1 \\ a_2 & d_2 & c_2 \\ a_3 & d_3 & c_3 \end{vmatrix}}{\begin{vmatrix} a_1 & b_1 & c_1 \\ a_2 & b_2 & c_2 \\ a_3 & b_3 & c_3 \end{vmatrix}}$$

And

$$z = \frac{\Delta z}{\Delta} = \frac{\begin{vmatrix} d_1 & b_1 & d_1 \\ d_2 & b_2 & d_2 \\ d_3 & b_3 & d_3 \end{vmatrix}}{\begin{vmatrix} a_1 & b_1 & c_1 \\ a_2 & b_2 & c_2 \\ a_3 & b_3 & c_3 \end{vmatrix}}$$

16.2. Inverse Matrix Method

This method based mainly on finding the inverse of the coefficient matrix [62].

This requires necessary and sufficient condition that is the coefficient matrix should be square matrix.

Let

$$Ax = b$$

By multiplying both sides of the equation by the inverse of the matrix coefficients, leads to:

$$A^{-1}Ax = A^{-1}b$$

It is well known that the multiplication of a matrix and its inverse leads to unit matrix, then:

$$Ix = A^{-1}b$$

Then

$$x = A^{-1}b$$

The inverse of the matrix can be found from:

$$A^{-1} = \frac{adj(A)}{\det(A)}$$

Now, it is important to remember the following restriction to solve system of linear equations using the inverse matrix method is that the determinant of the coefficient matrix not equals to zero.

If the determinant of the matrix coefficient equals zero, then we have two possibilities, the first one is the trival solution, and the second one is that we will have infinite number of solution.

Example

Solve, with the help of matrices, the following system of equations

$$3x + y + 2z = 3$$
$$2x - 3y - z = -3$$
$$x + 2y + z = 4$$

Solution

To understand solution procedure, let us explain how solution step with symbolic notations, so as one can understand the solution. After that numerical calculations with be done on a separate example.

Re-write the determinant coefficient as follows:

$$\Delta = \begin{vmatrix} a_1 & b_1 & c_1 \\ a_2 & b_2 & c_2 \\ a_3 & b_3 & c_3 \end{vmatrix}$$

We also knew before, how to evaluate the determinant, one can use Laplacian expansion, and this will lead to $\Delta = 8$

Next step, is to evaluate the matrix inversion, so as we can find the required solution from $x = A^{-1}b$.

$$\begin{vmatrix} x \\ y \\ z \end{vmatrix} = \frac{1}{\Delta} \begin{vmatrix} A_1 & A_2 & A_3 \\ B_1 & B_2 & B_3 \\ C_1 & C_2 & C_3 \end{vmatrix} \begin{vmatrix} d_1 \\ d_2 \\ d_3 \end{vmatrix}$$

Then

$$\begin{vmatrix} x \\ y \\ z \end{vmatrix} = \frac{1}{8} \begin{vmatrix} -1 & 3 & 5 \\ -3 & 1 & 7 \\ 7 & -5 & -11 \end{vmatrix} \begin{vmatrix} 3 \\ -3 \\ 4 \end{vmatrix} = \begin{vmatrix} 1 \\ 2 \\ -1 \end{vmatrix}$$

The final solution takes the following form:

$$\begin{vmatrix} x \\ y \\ z \end{vmatrix} = \begin{vmatrix} 1 \\ 2 \\ -1 \end{vmatrix}$$

Example

Solve, the following system of equations

$$2x + 3y - 4z = -3$$
$$3x - 2y + 5z = 2$$
$$x + 4y - 3z = -6$$

Solution

Starting the solution by finding the determinant of the coefficient matrix:

$$\Delta A = \begin{vmatrix} +2 & +3 & -4 \\ +3 & -2 & +5 \\ +1 & +4 & -3 \end{vmatrix} \begin{vmatrix} +2 & +3 \\ +3 & -2 \\ +1 & +4 \end{vmatrix} = -42$$

Write down the inverse formula:

$$A^{-1} = \frac{adj(A)}{\det(A)}$$

Next, the cofactor matrix takes the following form:

$$CM = \begin{pmatrix} +\begin{vmatrix} -2 & +5 \\ +4 & -3 \end{vmatrix} & -\begin{vmatrix} +3 & +5 \\ +1 & -3 \end{vmatrix} & +\begin{vmatrix} +3 & -2 \\ +1 & +4 \end{vmatrix} \\ -\begin{vmatrix} +3 & -4 \\ +4 & -3 \end{vmatrix} & +\begin{vmatrix} +2 & -4 \\ +1 & -3 \end{vmatrix} & -\begin{vmatrix} +2 & +3 \\ +1 & +4 \end{vmatrix} \\ +\begin{vmatrix} +3 & -4 \\ -2 & +5 \end{vmatrix} & -\begin{vmatrix} +2 & -4 \\ +3 & +5 \end{vmatrix} & +\begin{vmatrix} +2 & +3 \\ +3 & -2 \end{vmatrix} \end{pmatrix}$$

$$adj(A) = (C.M.)^T = \begin{pmatrix} -14 & -7 & +7 \\ +14 & -2 & -22 \\ +14 & -5 & -13 \end{pmatrix}$$

Then

$$A^{-1} = \frac{adj(A)}{\det(A)} = \left(\frac{1}{-42} \right) \begin{pmatrix} -14 & -7 & +7 \\ +14 & -2 & -22 \\ +14 & -5 & -13 \end{pmatrix}$$

$$= \begin{pmatrix} +\dfrac{1}{3} & +\dfrac{1}{6} & -\dfrac{1}{6} \\ -\dfrac{1}{3} & +\dfrac{1}{21} & +\dfrac{11}{21} \\ -\dfrac{1}{3} & +\dfrac{5}{42} & +\dfrac{13}{42} \end{pmatrix}$$

Then, the final solution takes the following form:

$$x = A^{-1}b = \begin{pmatrix} +\dfrac{1}{3} & +\dfrac{1}{6} & -\dfrac{1}{6} \\ -\dfrac{1}{3} & +\dfrac{1}{21} & +\dfrac{11}{21} \\ -\dfrac{1}{3} & +\dfrac{5}{42} & +\dfrac{13}{42} \end{pmatrix} \begin{pmatrix} -3 \\ +2 \\ -6 \end{pmatrix} = \begin{pmatrix} +\dfrac{1}{3} \\ -\dfrac{43}{21} \\ -\dfrac{13}{213} \end{pmatrix}$$

In the next two sections, two direct numerical methods will be presented in some details with illustrative examples. The first one is Gauss elimination method [63] and the second one is the modification to Gauss method, and is called Gauss-Jordan method [64].

16.3. Gauss Elimination Method

16.3.1. Main idea of Gauss Method

1. Construct the augmented matrix that is formed by putting the absolute terms with coefficient matrix in one matrix.

2. By some mathematical manipulation, and after so many operations, let us arrive to the limit that all the diagonal elements will be unity and all lower diagonal elements will be zero.

3. Third and final step, the back substitution one can get the required solution.

Example

Solve the following linear of equations

$$4x_1 + 3x_2 + 5x_3 = 3$$
$$2x_1 + x_2 + 3x_3 = 1$$
$$x_1 + 5x_2 + 6x_3 = 4$$

Solution

Form the augmented matrix, this matrix is formed by the coefficient matrix as well as absolute terms,

Formation the augmented matrix

$$\begin{pmatrix} 4 & 3 & 5 & 3 \\ 2 & 1 & 3 & 1 \\ 1 & 5 & 6 & 4 \end{pmatrix}$$

Divide elements of the 1st row by the element $a_{11} = 4$

$$\begin{pmatrix} 1 & \dfrac{3}{4} & \dfrac{5}{4} & \dfrac{3}{4} \\ 2 & 1 & 3 & 1 \\ 1 & 5 & 6 & 4 \end{pmatrix}$$

Multiply elements of the 2nd row by (-2), then add to the 1st row

$$\begin{pmatrix} 1 & \dfrac{3}{4} & \dfrac{5}{4} & \dfrac{3}{4} \\ 0 & -\dfrac{1}{2} & \dfrac{1}{2} & -\dfrac{1}{2} \\ 1 & 5 & 6 & 4 \end{pmatrix}$$

Multiply elements of the 3rd row by (-1), then add to the 1st row

$$\begin{pmatrix} 1 & \dfrac{3}{4} & \dfrac{5}{4} & \dfrac{3}{4} \\ 0 & -\dfrac{1}{2} & \dfrac{1}{2} & -\dfrac{1}{2} \\ 0 & -\dfrac{17}{4} & -\dfrac{19}{4} & -\dfrac{13}{4} \end{pmatrix}$$

Divide elements of the 2nd row by $-\left(\dfrac{1}{2}\right)$

$$\begin{pmatrix} 1 & \dfrac{3}{4} & \dfrac{5}{4} & \dfrac{3}{4} \\ 0 & 1 & -1 & 1 \\ 0 & -\dfrac{17}{4} & -\dfrac{19}{4} & -\dfrac{13}{4} \end{pmatrix}$$

Multiply elements of the 2nd row by $\left(\dfrac{17}{4}\right)$, then add to the 3rd row

$$\begin{pmatrix} 1 & \dfrac{3}{4} & \dfrac{5}{4} & \dfrac{3}{4} \\ 0 & 1 & -1 & 1 \\ 0 & 0 & 9 & -1 \end{pmatrix}$$

Divide the 3rd row by (9)

$$\begin{pmatrix} 1 & \dfrac{3}{4} & \dfrac{5}{4} & \dfrac{3}{4} \\ 0 & 1 & -1 & 1 \\ 0 & 0 & 1 & -\dfrac{1}{9} \end{pmatrix}$$

Back substitution step

To understand this step, let us re-write the system of equations, according to the last matrix obtained, as follows:

$$x_1 + \frac{3}{4}x_2 + \frac{5}{4}x_3 = \frac{3}{4}$$

$$x_2 - x_3 = 1$$

$$x_3 = -\frac{1}{9}$$

From the last equation, we get the value of x_3

$$x_3 = -\frac{1}{9}$$

$$x_2 - x_3 = 1 \Rightarrow x_2 = \frac{8}{9}$$

$$x_1 + \frac{3}{4}x_2 - \frac{5}{4}x_3 = \frac{3}{4} \Rightarrow x_1 = \frac{2}{9}$$

16.4. Gauss-Jordan Elimination Method

Gauss-Jordan elimination method is a modification of Gauss elimination method. The modification is that the upper and lower elements from the diagonal elements become zeroes and so there is no necessary to make the final step as in Gauss method, *i.e.,* there is no back substitution step to get the final results of the unknowns.

16.4.1. Main Idea of Gauss-Jordan Method

1. Construct the augmented matrix that is formed by putting the absolute terms with coefficient matrix in one matrix.

2. By some mathematical manipulation, and after so many operations, let us arrive to the limit that all the diagonal elements will be unity and all lower and upper diagonal elements will be zero.

From the last matrix, one can get the required results directly.

<u>Example</u>

Suppose you need to find numbers x_1, x_2 and x_3 such that the following three equations are all true:

$$2x_1 + 4x_2 + 2x_3 = 15$$
$$2x_1 + x_2 + 2x_3 = -5$$
$$4x_1 + x_2 - 2x_3 = 0$$

This is called a *system of linear equations* for the unknowns x_1, x_2 and x_3. The main idea of solution is as follows: eliminate x_1 from all but the first equation, eliminate x_2 from all but the second equation, and then eliminate x_3 from all but the third equation.

In the present example, we eliminate x_1 from the second equation by adding 3/2 times the first equation to the second, and then we eliminate x_1 from the third equation by adding the first equation to the third.

$2x + y - z = 8$

$.5y + .5z = 1$

$2y + z = 5$

Now we eliminate y from the first equation by adding -2 times the second equation to the first, and then we eliminate y from the third equation by adding -4 times the second equation to the third:

$2x - 2z = 6$

$0.5y + .5z = 1$

$- z = 1$

Finally, we eliminate z from the first equation by adding -2 times the third equation to the first, and then we eliminate z from the second equation by adding .5 times the third equation to the second:

$2x = 4$

$0.5y = 1.5$

$- z = 1$

The solution is:

$x = 2, y = 3$ and $z = -1$

Example

Solve the following linear of equations

$$4x_1 + 3x_2 + 5x_3 = 3$$
$$2x_1 + x_2 + 3x_3 = 1$$
$$x_1 + 5x_2 + 6x_3 = 4$$

Solution

Form the augmented matrix, this matrix is formed by the coefficient matrix as well as absolute terms,

$$\begin{pmatrix} 4 & 3 & 5 & 3 \\ 2 & 1 & 3 & 1 \\ 1 & 5 & 6 & 4 \end{pmatrix}$$

Divide elements of the 1st row by the element $a_{11} = 4$

$$\begin{pmatrix} 1 & \dfrac{3}{4} & \dfrac{5}{4} & \dfrac{3}{4} \\ 2 & 1 & 3 & 1 \\ 1 & 5 & 6 & 4 \end{pmatrix}$$

As done in the previous example, we will follow up the same procedure to make the lower diagonal elements equal zero. Start by multiplying the elements of the 2nd row by -2, then add the results to the elements of the 1st row, leads to:

$$\begin{pmatrix} 1 & \dfrac{3}{4} & \dfrac{5}{4} & \dfrac{3}{4} \\ 0 & -\dfrac{1}{2} & \dfrac{1}{2} & -\dfrac{1}{2} \\ 1 & 5 & 6 & 4 \end{pmatrix}$$

Multiply elements of the 3rd row by (-1), then add to the 1st row

$$\begin{pmatrix} 1 & \dfrac{3}{4} & \dfrac{5}{4} & \dfrac{3}{4} \\ 0 & -\dfrac{1}{2} & \dfrac{1}{2} & -\dfrac{1}{2} \\ 0 & -\dfrac{17}{4} & -\dfrac{19}{4} & -\dfrac{13}{4} \end{pmatrix}$$

Divide elements of the 2nd row by $-\left(\dfrac{1}{2}\right)$

$$\begin{pmatrix} 1 & \dfrac{3}{4} & \dfrac{5}{4} & \dfrac{3}{4} \\ 0 & 1 & -1 & 1 \\ 0 & -\dfrac{17}{4} & -\dfrac{19}{4} & -\dfrac{13}{4} \end{pmatrix}$$

The following two operations, will do simultaneously, multiply the elements of the 2nd row by $\dfrac{17}{4}$, then add to the corresponding elements of the 3rd row.

Multiply the elements of the 2nd row by $-\dfrac{3}{4}$, then add to the 1st row, the final result of these two operations is as follows:

$$\begin{pmatrix} 1 & 0 & 2 & 0 \\ 0 & 1 & -1 & 1 \\ 0 & 0 & 9 & -1 \end{pmatrix}$$

Divide the 3rd row by (9)

$$\begin{pmatrix} 1 & 0 & 2 & 0 \\ 0 & 1 & -1 & 1 \\ 0 & 0 & 1 & -\dfrac{1}{9} \end{pmatrix}$$

The following two operations, will do simultaneously, multiply the elements of the 3rd row by -2, then add to the corresponding elements of the 1st row.

Multiply the elements of the 3rd row by (1), then add to the 2nd row, the final result of these two operations is as follows:

$$\begin{pmatrix} 1 & 0 & 0 & \dfrac{2}{9} \\ 0 & 1 & 0 & \dfrac{8}{9} \\ 0 & 0 & 1 & -\dfrac{1}{9} \end{pmatrix}$$

As we see the coefficient of the 1st element in the 1st row equals unity and other coefficients are zeroes and the coefficient of the 2nd element in the 2nd row equals unity and other coefficients are zeroes and last the same for the 3rd row, we can get the direct values for the unknowns.

$$x_3 = -\frac{1}{9}$$

$$x_2 = \frac{8}{9}$$

$$x_1 = \frac{2}{9}$$

16.5. Jacobi Method

The present method differs from the above two methods, because it is an iterative method. This method requires very important condition that is the absolute value of the diagonal element in each row must be greater than or equal to the sum of the absolute values of all other elements in the same row [65].

Let us now show how to solve a linear system of linear algebraic equations using Jacobi's method. Assume that we want to solve the following system:

$$a_{11}x_1 + a_{12}x_2 + a_{13}x_3 + \ldots\ldots\ldots\ldots\ldots\ldots\ldots + a_{1n}x_n = b_1 \quad \text{(1)}$$

$$a_{21}x_1 + a_{22}x_2 + a_{23}x_3 + \ldots\ldots\ldots\ldots\ldots\ldots\ldots + a_{2n}x_n = b_2 \quad \text{(2)}$$

$$a_{31}x_1 + a_{32}x_2 + a_{33}x_3 + \ldots\ldots\ldots\ldots\ldots\ldots\ldots + a_{3n}x_n = b_3 \qquad (3)$$

$$\ldots\ldots\ldots\ldots\ldots\ldots\ldots\ldots$$

$$a_{n1}x_1 + a_{n2}x_2 + a_{n3}x_3 + \ldots\ldots\ldots\ldots\ldots\ldots\ldots + a_{nn}x_n = b_n \qquad (4)$$

The necessary condition will be as follows:

$$|a_{11}| \geq |a_{12}| + |a_{13}| + \ldots\ldots\ldots\ldots\ldots\ldots + |a_{1n}| \qquad (5)$$

$$|a_{22}| \geq |a_{21}| + |a_{23}| + \ldots\ldots\ldots\ldots\ldots\ldots + |a_{2n}| \qquad (6)$$

$$|a_{33}| \geq |a_{31}| + |a_{32}| + \ldots\ldots\ldots\ldots\ldots\ldots + |a_{3n}| \qquad (7)$$

$$\ldots\ldots\ldots\ldots\ldots\ldots$$

$$|a_{nn}| \geq |a_{n1}| + |a_{n2}| + \ldots\ldots\ldots\ldots\ldots\ldots + |a_{n,n-1}| \qquad (8)$$

The next step is to re-write the system in the following manner:

$$x_1 = \frac{b_1}{a_{11}} - \left(\frac{a_{12}}{a_{11}} x_2 + \frac{a_{13}}{a_{11}} x_3 + \ldots\ldots\ldots\ldots + \frac{a_{1n}}{a_{11}} x_n \right) \qquad (9)$$

$$x_2 = \frac{b_2}{a_{22}} - \left(\frac{a_{21}}{a_{22}} x_1 + \frac{a_{23}}{a_{22}} x_3 + \ldots\ldots\ldots + \frac{a_{2n}}{a_{22}} x_n \right) \qquad (10)$$

$$\ldots\ldots\ldots\ldots\ldots\ldots\ldots$$

$$x_n = \frac{b_n}{a_{nn}} - \left(\frac{a_{n1}}{a_{nn}} x_1 + \frac{a_{n2}}{a_{nn}} x_2 + \frac{a_{n3}}{a_{nn}} x_3 + \ldots + \frac{a_{n-1,n}}{a_{nn}} x_3 \right) \qquad (11)$$

This method also required initial values for all variables, then substitute by these initial values to get the new values of the same variables, and repeat this procedure to get the solution with the required accuracy.

Example

Using Jacob's method to solve the following system of equations:

$$-2x_1 + 5x_2 + x_3 - x_4 = -25$$
$$-x_1 - 2x_2 + 5x_3 + x_4 = 850$$
$$5x_1 + x_2 - x_3 - 2x_4 = -275 \qquad \textbf{(1)}$$
$$x_1 - x_2 - 2x_3 + 5x_4 = 725$$

Re-cast the given system as follows:

$$5x_1 + x_2 - x_3 - 2x_4 = -275$$
$$-2x_1 + 5x_2 + x_3 - x_4 = -25$$
$$-x_1 - 2x_2 + 5x_3 + x_4 = 850 \qquad \textbf{(2)}$$
$$x_1 - x_2 - 2x_3 + 5x_4 = 725$$

Re-cast the system given by equation (2) leads to:

$$x_1 + = \frac{1}{5}\left[-275 - \left(x_2 - x_3 - 2x_4\right)\right]$$

$$x_2 = \frac{1}{5}\left[-25 - \left(-2x_1 + x_3 - x_4\right)\right]$$

$$x_3 = \frac{1}{5}\left[850 - \left(-x_1 - 2x_2 + x_4\right)\right] \qquad \textbf{(3)}$$

$$x_4 = \frac{1}{5}\left[725 - \left(x_1 - x_2 - 2x_3\right)\right]$$

Up to this stage, we are ready to start computation, and using the initial values, as follows:

Let initial guess be $\left(x_1^0, x_2^0, x_3^0, x_4^0\right) = (0,0,0,0)$

Substitute by the given initial values $\left(x_1^0, x_2^0, x_3^0, x_4^0\right) = (0,0,0,0)$, in the system given by equation (3), to get the first approximation:

$$x_1^1 + = \frac{1}{5}[-275 - (0-0-0)] = \frac{-275}{5}$$

$$x_2 = \frac{1}{5}[-25 - (0+0-0)] = \frac{-25}{5}$$

$$x_3 = \frac{1}{5}[850 - (0-0+0)] = \frac{850}{5}$$ (4)

$$x_4 = \frac{1}{5}[725 - (0-0-0)] = \frac{725}{5}$$

Then the first approximation will be:

$$(x_1^1, x_2^1, x_3^1, x_4^1) = (-55, -5, 170, 145)$$

Fellow up the same procedure, one can get the second approximation as follows:

$$(x_1^2, x_2^2, x_3^2, x_4^2) = (38, -32, 128, 223)$$

$$(x_1^3, x_2^3, x_3^3, x_4^3) = (66.2, 29.2, 120.2, 183.2)$$

$$(x_1^4, x_2^4, x_3^4, x_4^4) = (47.76, 33.88, 158.48, 185.68)$$

$$(x_1^5, x_2^5, x_3^5, x_4^5) = (44.192, 19.544, 155.968, 205.616)$$

It is clear that we still need more iteration, because there exist errors between any two successive values of any of the four unknowns.

16.4. LU-Decomposition -Cholesky – Method

In this method, the coefficient matrix is analyzed to be the product of two matrices, they are L and U, where L is called lower triangle matrix and U is called the upper triangle matrix [66].

16.4.1. Steps of Solutions

1-We have $Ax = b$, let us assume $A = LU$, where L is the lower triangle matrix, while U is the upper triangle matrix,

2-Let $Ux = Y$, then Y can be found easily,

3-Then, solve $Ux = Y$

Let us show the reader, how LU-decomposition can be incorporated *via* a simple system of four linear equations as follows:

$$\begin{pmatrix} a_{11} & a_{12} & a_{13} & a_{14} \\ a_{21} & a_{22} & a_{23} & a_{24} \\ a_{31} & a_{32} & a_{33} & a_{34} \\ a_{41} & a_{42} & a_{43} & a_{44} \end{pmatrix} \begin{pmatrix} x_1 \\ x_2 \\ x_3 \\ x_4 \end{pmatrix} = \begin{pmatrix} b_1 \\ b_2 \\ b_3 \\ b_4 \end{pmatrix}$$

We assume $A = LU$, therefore, one can write in a matrix form the following:

$$\begin{pmatrix} a_{11} & a_{12} & a_{13} & a_{14} \\ a_{21} & a_{22} & a_{23} & a_{24} \\ a_{31} & a_{32} & a_{33} & a_{34} \\ a_{41} & a_{42} & a_{43} & a_{44} \end{pmatrix} = \begin{pmatrix} \ell_{11} & 0 & 0 & 0 \\ \ell_{21} & \ell_{22} & 0 & 0 \\ \ell_{31} & \ell_{32} & \ell_{33} & 0 \\ \ell_{41} & \ell_{42} & \ell_{43} & \ell_{44} \end{pmatrix} \times \begin{pmatrix} 1 & u_{12} & u_{13} & u_{14} \\ 0 & 1 & u_{23} & u_{24} \\ 0 & 0 & 1 & u_{34} \\ 0 & 0 & 0 & 1 \end{pmatrix}$$

Carry out the multiplication procedure to the right hand side and equate the correspondence with the left hand side, leads to:

$$\ell_{11} = a_{11}$$
$$\ell_{21} = a_{21}$$
$$\ell_{31} = a_{31}$$
$$\ell_{41} = a_{41}$$

$$\ell_{11} \times u_{12} = a_{12} \Rightarrow u_{12} = \frac{a_{12}}{\ell_{11}}$$

$$\ell_{11} \times u_{13} = a_{13} \Rightarrow u_{13} = \frac{a_{13}}{\ell_{11}}$$

$$\ell_{11} \times u_{14} \Rightarrow u_{14} = \frac{a_{14}}{\ell_{11}}$$

Also

$$\ell_{21} \times u_{12} + \ell_{22} = a_{22} \Rightarrow \ell_{22} = a_{22} - \ell_{21} \times u_{12}$$
$$\ell_{31} \times u_{12} + \ell_{32} = a_{23} \Rightarrow \ell_{32} = a_{23} - \ell_{31} \times u_{12}$$
$$\ell_{41} \times u_{12} + \ell_{42} = a_{42} \Rightarrow \ell_{42} = a_{42} - \ell_{41} \times u_{12}$$

And so on till we get:

$$u_{23} = \frac{a_{23} - \ell_{21} u_{13}}{\ell_{22}}$$

&

$$u_{24} = \frac{a_{24} - \ell_{21} u_{14}}{\ell_{22}}$$

One can generalize the above procedure, so as one can get the following general formulae:

$$\ell_{ij} = a_{ij} - \prod_{k=1}^{j-1} \ell_{ik} u_{kj}$$

&

$$u_{ij} = \frac{a_{ij} - \prod_{k=1}^{j-1} \ell_{ik} u_{kj}}{\ell_{ii}}$$

&

$$i, j = 1, 2, 3, \ldots, n$$

Example

Using LU-decomposition method to solve the following system of linear equations:

$$3x_1 - x_2 + 2x_3 = 12$$
$$x_1 + 2x_2 + 3x_3 = 11$$
$$2x_1 - 2x_2 - x_3 = 2$$

Solution

The first step is to write the coefficients matrix as follows:

$$\begin{pmatrix} 3 & -1 & 2 \\ 1 & 2 & 3 \\ 2 & -2 & -1 \end{pmatrix} = \begin{pmatrix} \ell_{11} & 0 & 0 \\ \ell_{21} & \ell_{22} & 0 \\ \ell_{31} & \ell_{32} & \ell_{33} \end{pmatrix} \times \begin{pmatrix} 1 & u_{12} & u_{13} \\ 0 & 1 & u_{23} \\ 0 & 0 & 1 \end{pmatrix}$$

Carry out the procedure explained before, one can get:

$$\begin{pmatrix} \ell_{11} & 0 & 0 \\ \ell_{21} & \ell_{22} & 0 \\ \ell_{31} & \ell_{32} & \ell_{33} \end{pmatrix} = \begin{pmatrix} 3 & 0 & 0 \\ 1 & \dfrac{7}{3} & 0 \\ 2 & -\dfrac{4}{3} & 1 \end{pmatrix}$$

$$\begin{pmatrix} 1 & u_{12} & u_{13} \\ 0 & 1 & u_{23} \\ 0 & 0 & 1 \end{pmatrix} = \begin{pmatrix} 1 & -\dfrac{1}{3} & \dfrac{1}{3} \\ 0 & 1 & 1 \\ 0 & 0 & 1 \end{pmatrix}$$

Now

$$\begin{pmatrix} 3 & 0 & 0 \\ 1 & \dfrac{7}{3} & 0 \\ 2 & -\dfrac{4}{3} & 1 \end{pmatrix} \begin{pmatrix} y_1 \\ y_2 \\ y_3 \end{pmatrix} = \begin{pmatrix} 12 \\ 11 \\ 2 \end{pmatrix}$$

$$\Rightarrow$$

$$y_1 = 4 \quad \& \quad y_2 = 3 \quad \& \quad y_3 = 2$$

$$\Rightarrow$$

$$x_1 = 3 \quad \& \quad x_2 = 1 \quad \& \quad x_3 = 2$$

SUPPLEMENTARY PROBLEMS

Problem (1)

Write down for a system of n linear equations and of n unknowns the different forms that can represent this system?

Problem (2)

The equations $5x + 2y = 48$ and $3x + 2y = 32$ represent the money collected from school concert tickets sales during two class periods.

If x represents the cost for each adult ticket and y represents the cost for each student ticket, what is the cost for each adult ticket?

Problem (3)

Margie is responsible for buying a week's supply of food and medication for the dogs and cats at a local shelter. The food and medication for each dog costs twice as much as those supplies for a cat. She needs to feed 164 cats and 24 dogs. Her budget is $4240.

How much can Margie spend on each dog for food and medication?

Problem (4)

Solve and check

$$x + y = 1$$
$$x + 3y = 9$$

Problem (5)

For the following system of linear equations:

$$-2x_1 + 5x_2 + x_3 = 125$$
$$-x_1 - 2x_2 + 5x_3 = 250$$
$$5x_1 + x_2 - x_3 = 275$$

Write down in some details the necessary and sufficient condition to use the matrix inversion method to solve such a system?

Problem (6)

Using the inverse matrix method to find the solution of the following system of linear equations:

$$-2x_1 + 5x_2 + x_3 = 125$$
$$-x_1 - 2x_2 + 5x_3 = 250$$
$$5x_1 + x_2 - x_3 = 275$$

Problem (7)

Use the Gauss elimination method to solve the following linear of equations

$$8x_1 + 6x_2 + 10x_3 = 6$$
$$4x_1 + 2x_2 + 6x_3 = 2$$
$$2x_1 + 10x_2 + 12x_3 = 8$$

Problem (8)

Use the Gauss-Jordan elimination method to solve the following linear of equations

$$8x_1 + 6x_2 + 10x_3 = 6$$
$$4x_1 + 2x_2 + 6x_3 = 2$$
$$2x_1 + 10x_2 + 12x_3 = 8$$

Problem (9)

Use the LU-decomposition method to solve the following linear of equations

$$8x_1 + 6x_2 + 10x_3 = 6$$
$$4x_1 + 2x_2 + 6x_3 = 2$$
$$2x_1 + 10x_2 + 12x_3 = 8$$

Problem (10)

Compare the results obtained in problems (4-6) and give and explanation between the minor or major differences in the accuracy obtained.

CHAPTER 4

Numerical Solution of Nonlinear Equations

Abstract: In the present chapter, we will focus on the methods of solution of nonlinear single and system of nonlinear equations. The methods presented herein are explained in brief details with error analysis and how the convergence and the stability occurred.

Keywords: Bisection, Convergence and stability analysis, Error analysis, Newton-Raphson, Simple iteration.

1. INTRODUCTION

The main object of the present chapter is to understand well how to solve non-linear equation and non-linear system of equations [69].

Some famous methods for solving non-linear equation and equations will be discussed with illustrated examples provided. There are closed form solutions for quadratic and even 3rd degree polynomial equations.

Higher degree polynomials can sometimes be factored. However, in general there is no closed form analytical solution to non-linear equations.

Let us start these methods by the famous simple iterative method, then followed by other different methods with illustrative examples.

2. ITERATIVE METHODS

2.1. Simple Iteration Method

The simple iteration method is an iterative method to find the roots of a given non-linear equation [70]. Simply this method provides acceptable results but to do that, some conditions should be satisfied in advance to make sure that we will get the required solution with acceptable accuracy and less iteration numbers.

2.1.1. Summary of Simple Iteration Method

To summarize the main steps of the solution procedure, let us have a non-linear equation of the form $f(x) = 0$ and it is required to solve this equation, *i.e.*, to find its roots, to do that, let us extract another form from the original one but this form should satisfy some condition. The extracted form will take the following form:

Said Gamil Ahmed Sayed Ahmed

$$x = \Phi(x)$$

Now, then the variable in the left hand-side will be the required root and the variable inside the function $\Phi(x)$ will be the iterated value at previous step, therefore, let us write the extracted form as follows:

$$x_{n+1} = \Phi\left(x_n\right)$$

Next step is to write the necessary condition before starting the iterations that will be:

$$\left|\Phi\left(x_n\right)\right| \le 1$$

This condition should be satisfied for any interval containing the required root.

Example

Use simple iteration method to solve $\sin x - x + 0.25 = 0$. Choose $x_0 = 1.2$ as initial starting point.

Solution

The first step as mentioned before is to re-write the given equation in a suitable form $x = \Phi(x)$, such that the necessary condition should be satisfied.

It is also important to remember that one can obtain different forms from the given equation, and he should take into consideration that the best choice is that which satisfies the necessary condition.

Due to the above explanation, one can get two possible forms, they are:

Form (1): $x = \sin x + 0.25$

Form (2): $x = \sin^{-1}(0.25 - x)$

Let us now start checking the previous two forms and we will check the first form.

Therefore, let us re-write the first form as follows:

$$x_{n+1} = \Phi(x) = \left(\sin x_n + 0.25 \right) \tag{1}$$

Let us find the first derivative, leads to:

$$\Phi'(x_n) = \cos x_n \tag{2}$$

The next step is to determine the interval containing the root, and this can be done by successive substitution in the given equation by an integer values, and focus on the resulted sign not the number itself, and when two different signs obtained in two successive integer numbers this will be the interval containing the required root.

$$x_{n+1} = \Phi(x_n) = \left[\sin x_n + 0.25 \right], n = 0,1,2,3,4,5,\dots$$

Staring with n = 0

$$x_1 = \sin 1.2 + 0.25 = 1.182$$

$$x_2 = \sin 1.182 + 0.25 = 1.17$$

$$x_3 = \sin 1.17 + 0.25 = 1.173$$

$$x_4 = 1.172$$

$$x_5 = 1.172$$

It is clear the root in the fourth and fifth iterations are the same for three decimal places, therefore, if the required root required correct to three decimal places.

Then we can stop at this iteration, but if more accuracy is required then, the number of iterations should be increased till the required accuracy achieved.

2.2. Newton-Raphson Method

The idea of the method is as follows: one starts with an initial guess which is reasonably close to the true root, then the function is approximated by its tangent line and one computes the x-intercept of this tangent line.

This x-intercept will typically be a better approximation to the function's root than the original guess, and the method can be iterated.

Let us derive the general formula for Newton's iterative formula using the concept Taylor's series [71]:

$$f(x) = f(x_o) + (x - x_o)f'(x_o) + (x - x_o)^2 \frac{f''(x_o)}{2!} + (x - x_o)^3 \frac{f'''(x_o)}{3!} + ..$$ Let's say

that x is "close" to x_o and keep just the first two terms.

$$f(x) \approx f(x_o) + (x - x_o)f'(x_o)$$

We want to solve for x such that $f(x) = 0$.

$$f(x_o) + (x - x_o)f'(x_o) = 0$$

$$x = x_o - \frac{f(x_o)}{f'(x_o)}$$

In effect we have approximated $f(x)$ by a straight line; x is the intercept of that line with the x-axis. It may or may not be a good approximation for the root α.

We know from the definition of the derivative at a given point that it is the slope of a tangent at that point, that is:

$$f'(x_n) = \frac{\Delta y}{\Delta x} = \frac{f(x_n)}{x_n - x_{n+1}}$$

Here, f' denotes the derivative of the function f, then:

$$x_{n+1} = x_n - \frac{f(x_n)}{f'(x_n)}$$

We start the process off with some arbitrary initial value x_0. The method will usually converge, provided this initial guess is close enough to the unknown zero, and that $f'(x_o) \neq 0$.

Example

Evaluate the cube root of 467.

Solution

We want to find a value of x such that $x^3 = 467$.

Put another way, we want to find a *root* of the following equation:

$$f(x) = x^3 - 467 = 0.$$

If we take our *initial guess* to be $x_o = 6$, then by *iterating* the formula above, we generate the following results:

$$x_1 \cong x_o - \frac{f(x_o)}{f'(x_{1o})}$$

$$= 6 - \frac{-251}{108}$$

$$= 8.32407$$

$$x_2 \cong x_1 - \frac{f(x_1)}{f'(x_1)}$$

$$= 8.32407 - \frac{109.7768}{207.8706}$$

$$= 7.79597$$

$$x_3 \cong x_2 - \frac{f(x_2)}{f'(x_2)}$$

$$= 7.79597 - \frac{6.817273}{182.33156}$$

$$= 7.75858$$

Example

Consider the problem of finding the positive number x with $\cos x = x^3$.

Solution

Re-write the equation as follows:

$$f(x) = \cos x - x^3 = 0$$

Then

$$f'(x) = -\sin x - 3x^2 = 0$$

Since, $\cos x \leq 1$ for all x and $x^3 > 1$ for $x > 1$, and we know that our zero lies between 0 and 1. Therefore, let us take the initial value $x_0 = 0.5$

The iterative procedure will be as follows;

$$x_1 = x_0 - \frac{f(x_0)}{f'(x_0)}$$
$$= 0.5 - \frac{\cos(0.5) - (0.5)^3}{-\sin(0.5) - 3 \times (0.5)^3}$$
$$= 1.112141637097$$

$$x_2 = x_1 - \frac{f(x_1)}{f'(x_1)}$$
$$= 0.909672693736$$

$$x_3 = x_2 - \frac{f(x_2)}{f'(x_2)}$$
$$= 0.867263818209$$

$$x_4 = x_3 - \frac{f(x_3)}{f'(x_3)}$$
$$= 0.865477135298$$

$$x_5 = x_4 - \frac{f(x_4)}{f'(x_4)}$$
$$= 0.865474033111$$

We see that the number of correct digits after the decimal point increases from 2 (for x_3) to 5 and 10, illustrating the quadratic convergence.

Example

Apply the Newton-Raphson method to the function:

$$f(x) = x^3 - 3 .$$

Solution

$$f(1) = (1)^3 - 3 = -ve$$
$$f(2) = (2)^3 - 3 = +ve$$

The iterations are given by the formula

$$x_{n+1} = x_n - \frac{f(x_n)}{f'(x_n)} = x_n - \frac{x_n^3 - 3}{3x_n^2}$$

$x_0 = 1$
$x_1 = 1.6666666666666666$
$x_2 = 1.4711111111111111$
$x_3 = 1.4428120982493343$
$x_4 = 1.4422497859899960$
$x_5 = 1.4422499570307441$

Example

Find an approximation to $\sqrt{5}$ to ten decimal places.

Solution

$$f(x) = x^2 - 5 = 0$$

$$f(1) = (1)^2 - 5 = -ve$$

$$f(2) = (2)^2 - 5 = -ve$$

$$f(3) = (3)^2 - 5 = +ve$$

$$x_{n+1} = x_n - \frac{f(x_n)}{f'(x_n)} = x_n - \frac{x_n^2 - 5}{3x_n}$$

$$x_0 = 2$$

$$x_1 = 2.25$$

$$x_2 = 2.23611111111111111$$

$$x_3 = 2.36067977915804002$$

$$x_4 = 2.23606797749978$$

$$x_5 = 2.23606797749978$$

2.3. Bisection Method

The bisection method is a root-finding iterative algorithm in which the interval containing the root is divided in half and then selects the subinterval in which a root exists [72]. Suppose we want to solve the equation $f(x) = 0$.

2.3.1. Solution Steps

1-Assume that the required root lies in the closed interval $[a, b]$ such that the sign of $f(a)$ is different from sign of $f(b)$.

2-Find the mid point c between a and b from $c = \dfrac{a+b}{2}$, then find $f(c)$

3-Have a look to the signs of $f(a)$, $f(b)$ and $f(c)$, then the root will be in between the two points having different signs of their function value, *i.e.*, if a and c will have $f(a)$ and $f(c)$ of different signs.

4-This means that the root will be between a and c, and so on, repeat the same procedure.

The following Fig. (1) shows the procedure clearly.

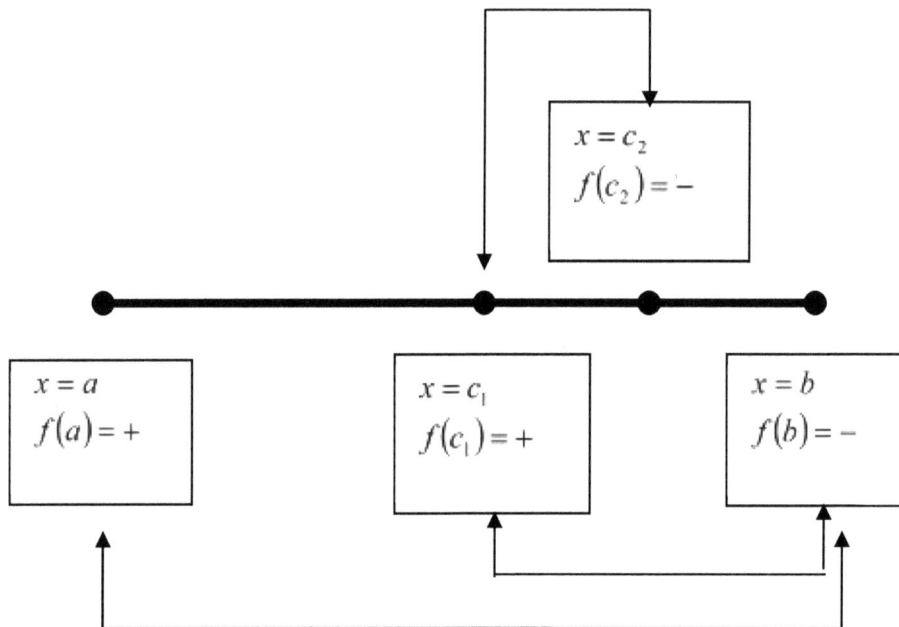

Fig. (1). Bisection graphically.

Example

Compute numerical approximations for $\sqrt{2}$.

Solution

$$f(x) = x^2 - 2 = 0$$

$$a = 1 \quad \& \quad f(1) = -1$$
$$b = 2 \quad \& \quad f(2) = +2$$
$$\Rightarrow c = \frac{1+2}{2} = 1.5$$
$$\&$$
$$f(1.5) = +0.25$$

Table 1. Example computations.

n	a_n	b_n
0	1	2
1	1	1.5
2	1.25	1.5
3	1.375	1.5
.....	.....	.....
20	1.414213181	1.414214134

3. SYSTEM OF NON-LINEAR EQUATIONS

In the present section, will focus on the famous methods for solving system of non-linear equations. We will start deriving the Newton-Raphson method, followed by the simple iteration method [73].

3.1. Newton-Raphson Method

The two simultaneous equations will have the form:

$$F(x, y) = 0$$
$$G(x, y) = 0 \tag{1}$$

We are looking for a point (x_0, y_0) that, represent the required root, to do that follow up the following procedure.

Let

$$(x_0 + e_1, y_0 + e_2)$$

Be the solution of the equation (1), therefore;

$$F(x_0 + e_1, y_0 + e_2) = 0$$

And

$$G(x_0 + e_1, y_0 + e_2) = 0$$

Using Taylor's expansion around the starting point (x_0, y_0), one can get:

$$F(x_0, y_0) + e_1\left(\frac{\partial F}{\partial x}\right)_0 + e_2\left(\frac{\partial F}{\partial y}\right)_0 = 0$$

$$G(x_0, y_0) + e_1\left(\frac{\partial G}{\partial x}\right)_0 + e_2\left(\frac{\partial G}{\partial y}\right)_0 = 0$$

The higher powers of e_1 and e_2 are neglected due to their very small values.

The above equations can be reduced to

$$e_1\left(\frac{\partial F}{\partial x}\right)_0 + e_2\left(\frac{\partial F}{\partial y}\right)_0 = -F_0$$

$$e_1\left(\frac{\partial G}{\partial x}\right)_0 + e_2\left(\frac{\partial G}{\partial y}\right)_0 = -G_0$$

(2)

Solving equation (2), we get:

$$\therefore e_1 = -\frac{\begin{vmatrix} F & F_y \\ G & G_y \end{vmatrix}_{(x_0, y_0)}}{\begin{vmatrix} F_x & F_y \\ G_x & G_y \end{vmatrix}_{(x_0, y_0)}}$$

And

$$e_2 = -\frac{\begin{vmatrix} F_x & F \\ G_x & G \end{vmatrix}_{(x_0, y_0)}}{\begin{vmatrix} F_x & F_y \\ G_x & G_y \end{vmatrix}_{(x_0, y_0)}}$$

Hence the first approximation to the required root is;

$$x_1 = x_0 + e_1 = x_0 - \frac{\begin{vmatrix} F & F_y \\ G & G_y \end{vmatrix}_{(x_0,y_0)}}{\begin{vmatrix} F_x & F_y \\ G_x & G_y \end{vmatrix}_{(x_0,y_0)}}$$

$$y_1 = y_0 + e_2 = y_0 - \frac{\begin{vmatrix} F_x & F \\ G_x & G \end{vmatrix}_{(x_0,y_0)}}{\begin{vmatrix} F_x & F_y \\ G_x & G_y \end{vmatrix}_{(x_0,y_0)}}$$

In general, one can write:

$$x_{n+1} = x_n - \frac{\begin{vmatrix} F & F_y \\ G & G_y \end{vmatrix}_{(x_n,y_n)}}{\begin{vmatrix} F_x & F_y \\ G_x & G_y \end{vmatrix}_{(x_n,y_n)}}$$

And

$$y_{n+1} = y_n - \frac{\begin{vmatrix} F_x & F \\ G_x & G \end{vmatrix}_{(x_n,y_n)}}{\begin{vmatrix} F_x & F_y \\ G_x & G_y \end{vmatrix}_{(x_n,y_n)}}$$

Example

Solve to get the root of the equations,

$x^2 + y = 3$, and $x + y^2 = 5$, Close to the point (2,-2)

Solution

Put the two equations in the form of homogenous, *i.e,*

$$x^2 + y - 3 = 0 = F(x, y)$$

And

$$x + y^2 - 5 = 0 = G(x, y)$$

$$F_x = \frac{\partial F}{\partial x} = 2x$$

$$F_y = \frac{\partial F}{\partial y} = 1,$$

$$G_x = \frac{\partial G}{\partial x} = 1$$

And

$$G_y = \frac{\partial G}{\partial y} = 2y$$

For $(x_0, y_0) = (2, -2)$

Then

$$x_{n+1} = x_n - \frac{\begin{vmatrix} x_n^2 + y_n - 3 & 1 \\ x_n + y_n^2 - 5 & 2y_n \end{vmatrix}}{\begin{vmatrix} 2x_n & 1 \\ 1 & 2y_n \end{vmatrix}}$$

$$x_{n+1} = \frac{2x_n^2 y_n - y_n^2 + 6y_n - 5}{4x_n y_n - 1}$$

$$y_{n+1} = y_n - \frac{\begin{vmatrix} 2x_n & x_n^2 + y_n - 3 \\ 1 & x_n + y_n^2 - 5 \end{vmatrix}}{\begin{vmatrix} 2x_n & 1 \\ 1 & 2y_n \end{vmatrix}}$$

$$y_{n+1} = \frac{2x_n y_n^2 - x_n^2 + 10x_n - 3}{4x_n y_n - 1}$$

Similarly,

$$x_2 = 2.1691432 \text{, and } y_2 = -1.68736$$

Then repeat the process till the required accuracy is achieved.

Example

Solve to get the root of the equations

$$x^2 + y^2 = 1$$

And

$$x^3 = y,$$

Correct to three decimals. Start solution with $x_0 = 0.8$, $y_0 = 0.55$

Solution

Put the two equations in the form of homogenous, *i.e.*

$$F(x, y) = x^2 + y^2 - 1 = 0$$

And

$$G(x, y) = x^3 - y = 0$$

$$F_x = \frac{\partial F}{\partial x} = 2x$$

$$F_y = \frac{\partial F}{\partial y} = 2y$$

$$G_x = \frac{\partial G}{\partial x} = 3x^2 \text{ And } G_y = \frac{\partial G}{\partial y} = -1$$

For $(x_0 = 0.8, y_0 = 0.55)$

$$x_{n+1} = x_n - \frac{\begin{vmatrix} x^2 + y^2 - 1 & 2y_n \\ x^3_n - y_n & -1 \end{vmatrix}}{\begin{vmatrix} 2x_n & 2y_n \\ 3x_n^2 & -1 \end{vmatrix}}$$

$$y_{n+1} = y_n - \frac{\begin{vmatrix} 2x_n & x^2 + y^2 - 1 \\ 3x_n^2 & x^3_n - y_n \end{vmatrix}}{\begin{vmatrix} 2x_n & 2y_n \\ 3x_n^2 & -1 \end{vmatrix}}$$

$$x_1 = 0.8 - \frac{\begin{vmatrix} -0.057 & 1.1 \\ -0.038 & -1 \end{vmatrix}}{\begin{vmatrix} 1.6 & 1.1 \\ 1.92 & -1 \end{vmatrix}} = 0.8267 \quad y_1 = 0.55 - \frac{\begin{vmatrix} 1.6 & -0.0575 \\ 1.92 & -0.088 \end{vmatrix}}{\begin{vmatrix} 1.6 & 1.1 \\ 1.92 & -1 \end{vmatrix}} = 0.56336$$

Example

Solve

$$x^2 + y = 3$$

And

$$x + y^2 = 5$$

Close to the point (2,-2)

Solution

$$F(x,y) = x^2 + y - 3 = 0$$
$$G(x,y) = x + y^2 - 5 = 0$$

$$F_x = \frac{\partial F}{\partial x} = 2x$$

$$F_y = \frac{\partial F}{\partial y} = 1$$

$$G_x = \frac{\partial G}{\partial x} = 1$$

$$G_y = \frac{\partial G}{\partial y} = 2y$$

$$x_{n+1} = x_n - \frac{\begin{vmatrix} x_n^2 + y_n - 3 & 1 \\ x_n + y_n^2 - 5 & 2y_n \end{vmatrix}}{\begin{vmatrix} 2x_n & 1 \\ 1 & 2y_n \end{vmatrix}}$$

$$x_{n+1} = \frac{2x_n^2 y_n - y_n^2 + 6y_n - 5}{4x_n y_n - 1}$$

$$y_{n+1} = y_n - \frac{\begin{vmatrix} 2x_n & x_n^2 + y_n - 3 \\ 1 & x_n + y_n^2 - 5 \end{vmatrix}}{\begin{vmatrix} 2x_n & 1 \\ 1 & 2y_n \end{vmatrix}}$$

$$y_{n+1} = \frac{2x_n y_n^2 - x_n^2 + 10x_n - 3}{4x_n y_n - 1}$$

Similarly

$$x_2 = 2.1691432$$

$$y_2 = -1.68736$$

Example

Solve to get the root of the equations

$$x^2 + y^2 = 1$$

And

$$x^3 = y$$

Correct to three decimals.

Start solution with $x_0 = 0.8$, $y_0 = 0.55$

Solution

$$F(x, y) = x^2 + y^2 - 1 = 0$$

$$G(x, y) = x^3 - y = 0$$

$$F_x = \frac{\partial F}{\partial x} = 2x \quad G_x = \frac{\partial G}{\partial x} = 3x^2$$

$$F_y = \frac{\partial F}{\partial y} = 2y \quad G_y = \frac{\partial G}{\partial y} = -1$$

$$x_{n+1} = x_n - \frac{\begin{vmatrix} x^2 + y^2 - 1 & 2y_n \\ x^3_n - y_n & -1 \end{vmatrix}}{\begin{vmatrix} 2x_n & 2y_n \\ 3x_n^2 & -1 \end{vmatrix}}$$

$$y_{n+1} = y_n - \frac{\begin{vmatrix} 2x_n & x^2 + y^2 - 1 \\ 3x^2_n & x^3_n - y_n \end{vmatrix}}{\begin{vmatrix} 2x_n & 2y_n \\ 3x_n^2 & -1 \end{vmatrix}}$$

$$x_1 = 0.8 - \frac{\begin{vmatrix} -0.057 & 1.1 \\ -0.038 & -1 \end{vmatrix}}{\begin{vmatrix} 1.6 & 1.1 \\ 1.92 & -1 \end{vmatrix}}$$

$$= 0.8267$$

$$y_1 = 0.55 - \frac{\begin{vmatrix} 1.6 & -0.0575 \\ 1.92 & -0.088 \end{vmatrix}}{\begin{vmatrix} 1.6 & 1.1 \\ 1.92 & -1 \end{vmatrix}}$$

$$= 0.56336$$

3.2. Simple Iteration Method

In this section, we will develop the simple iteration method to solve a system of two nonlinear equations, by the so called simple iteration method [74-77].

As described earlier when dealing with the same method but for a single nonlinear equation, we start the solution by extracting iterative formula of the form $x_{n+1} = \Phi(x_n)$ and before iterations.

This formula should be checked for convergency and stability by finding the first derivative, then check the convergence condition, that was $|\Phi(x_n)| \leq 1$, $\forall x \in [a, b]$.

Important notes should be achieved:

1- $|\Phi(x_n)| \leq 1$, $x = a$

2- $|\Phi(x_n)| \leq 1$, $x = b$

3-The first note should be achieved.

4-In some cases, the second note is not achieved and we have to decrease the upper interval limit and check again, till the condition in the second note should be achieved.

3.2.1. Solution Procedure

To summarize the main steps of the solution procedure, let us have a system of two non-linear equations of the form:

$$f(x, y) = 0$$

And

$$g(x, y) = 0$$

With initial starting point (x_0, y_0)

It is required to solve this system, *i.e.*, to find the point of intersection of the two equations.

To do that, let us extract other forms from the original ones, as follows:

$$f(x, y) = 0 \Rightarrow$$
$$x_{n+1} = \Phi(x_n, y_n)$$

Or

$$y_{n+1} = \Phi(x_n, y_n)$$

And

$$g(x, y) = 0 \Rightarrow$$
$$x_{n+1} = \Psi(x_n, y_n)$$

Or

$$y_{n+1} = \Psi(x_n, y_n)$$

Notes

1-If we choose $x_{n+1} = \Phi(x_n, y_n)$ from the first equation, then $y_{n+1} = \Psi(x_n, y_n)$ will be from the second one and vice versa.

2-If we choose $x_{n+1} = \Psi(x_n, y_n)$ from the first equation, then $y_{n+1} = \Phi(x_n, y_n)$ will be from the second one and *vice versa*.

3-Check convergence condition:

If we will use $x_{n+1} = \Phi(x_n, y_n)$ & $y_{n+1} = \Psi(x_n, y_n)$, then the condition will be:

$$\left| \Phi(x_n, y_n) \right|_{(x_0, y_0)} \leq 1,$$
$$\left| \Psi(x_n, y_n) \right|_{(x_0, y_0)} \leq 1$$

SUPPLEMENTARY PROBLEMS

Problem (1)

Use any numerical method to find the smallest positive root of the equation (correct to four decimal places),

$$2x^3 + 3x^2 - 3x - 3 = 0.0$$

Problem (2)

Use simple interaction method (4 interactions) to find the two intersections of the two curves $y_2=3x$, and $y_1= e^x$, near x= 0.6, (with initial guess $x_0= 0$ or $x_0=1$)

Problem (3)

Use simple iteration method to Find the root of the equation $4x-(x^3+1)=0$, to eight decimal places take $x_0= 0$.

Problem (4)

Use Newton method to Find the root of the equation $e^x - 3x^2 = 0$, to eight decimal places ($x_0= -0.5$)

Problem (5)

Solve the equation $f(x)=x^3-4x+1$ by Newton's method, three iterations are required (take $x_0= 0$).

Problem (6)

Use the initial values $x_0=2.4$ and $y_0=2.6$, to solve the two simultaneous equations. $x^2+(y-4)^2-9=0,(x-4)^2+y^2-9=0$, using Newton method.

Problem (7)

Solve the equation $f(x) = e^{-x} + x^2 -10=0$

by Newton's method, three iterations are required.

Problem (8)

For the system,

F (x,y) = 2x²-xy-5x+1 = 0

G (x,y) = x+3 log x-y² = 0,

Find the positive root to four significant digits. Given that (x₀,y₀)= (3.5, 2.2).

Problem (9)

Use Newton method to solve the two simultaneous equations:

$x^2 + (y-4)^2 - 9 = 0, (x-4)^2 + y^2 - 9 = 0$, Using the initial values $x_0 = 2.4$ and $y_0 = 2.6$.

(Two iterations are required)

Problem (10)

Solve the following linear of equations, using:

(1)   Gauss elimination method

(2)   Gauss Jordan method

$3x_1 - 5x_2 + 7x_3 = 2 \ 3_1 + 5x_2 + x_3 = 20$
$3x_1 + 2x_2 - 3x_3 = 9 \ 30x_1 + 12x_2 - 13x_3 = 19$
$2x_1 + 7x_2 + x_3 = -4 \ 21x_1 + 3x_2 + 2x_3 = -14$

Problem (11)

Use a numerical method to find the smallest positive root of the equation (correct to four decimal places) $2x^3 + 3x^2 - 3x - 3 = 0.0$

Problem (12)

Use simple interaction method (4 interactions) to find the two intersections of the two curves $y_1 = e^x$ and $y_2 = 3x$ near $x = 0.6$ and an initial guess $x_0 = 0$

Problem (13)

Use simple iteration method to Find the root of the equation $4x - x^3 - 1 = 0$ to three decimal places and $x_0 = 0$

Problem (14)

Use Newton method to find the root of the equation $e^x - 3x^2 = 0$ correct to eight decimal places $x_0 = -0.5$

Problem (15)

Solve the equation $f(x) = x^3 - 4x + 1$ using Newton's method up to three iterations with $x_0 = 0$

Problem (16)

Using Newton method and an initial value $x_0 = 2.4$ & $y_0 = 2.6$ to solve the following two simultaneous equations: $x^2 + (y - 4)^2 - 9 = 0$ and $(x - 4)^2 + y^2 - 9 = 0$

CHAPTER 5

Numerical Solution for Ordinary Differential Equations

Abstract: In science, engineering, and technology most of these applications contain one or two systems of ordinary differential equations. Many of these differential equations can't be solved using the traditional analytical method called symbolic methods. In these cases numerical or approximate methods will be used instead. In the present chapter, some elementary basic numerical methods for solving one, two and higher ordinary differential equations as well as system of ordinary differential equations will be presented with illustrative examples.

Keywords: Euler method, Modified Euler, Runge-Kutta method, Taylor series method.

1. INTRODUCTION

In science, engineering, and technology most of these applications contain one, two or more systems of ordinary differential equations. Many of these differential equations can't be solved using the traditional analytical methods called symbolic methods [78].

This chapter focuses on the analysis of basic and elementary numerical methods to solve one, two and higher ordinary differential equations as well as system of ordinary differential equations. There are three main types of solutions of any mathematical problem, analytical, approximate, and numerical and any combination between two or more methods of solution will take the name hybrid (mixed) methods.

The most important type is the analytical methods [79] but usually this type of solution is not so easy to obtain due to so many reasons. Some of these reasons are the nature of the problem underhand or due to nonlinearity behavior and more.

Approximate methods are developed in such a way that major part of the method is carried out analytically, and then numerical methods are used to find the final solutions. Numerical methods are wide since long time and became ever more comprehensive with the rapid development of computers, and finally the hybrid methods are combination of two or more numerical methods.

In the present chapter, some elementary basic numerical methods for solving one, two and higher ordinary differential equations as well as system of ordinary differential equations will be presented in brief details with illustrative examples.

2. TAYLOR SERIES METHOD FOR 1ST ORDER O.D.E.

In numerical differentiation, we learned how to use Taylor's series to evaluate the derivatives numerically. Refer back to the basic definition of the ordinary differential equation, is that equation involves variables and derivatives in a certain manner.

What we want to say here is that, Taylor's series is one but not the best method to solve the ordinary differential equation.

To understand how Taylor's series can be used to find the numerical solution of the ordinary differential equation, let's do it on a first order differential equation.

Assume that we have a first order differential equation of the following form:

$$y' = f(x, y) \tag{1}$$

With the following boundary condition:

$$y(x_0) = y_0 \tag{2}$$

The boundary condition is itself a solution for the given ordinary differential equation at a certain point on the boundary, and so it will be required to find such a solution at another point over the domain of the definition.

Now let us have a look at the Taylor's series about an expansion point $x = x_0$.

The Taylor's series expansion of a function $f(x)$ around a point of expansion x has the following form:

$$f(x) = f(x_0) + \frac{(x - x_0)}{1!} f'(x_0) + \frac{(x - x_0)^2}{2!} f''(x_0) + \ldots \ldots \tag{3}$$

Assume that the difference in distance between two successive points over the domain solution is defined as:

$$h = x - x_0 \qquad (4)$$

By making use of equation (4) into equation (3), we get:

$$f(x) = f(x_0) + \frac{h}{1!} f'(x_0) + \frac{h^2}{2!} f''(x_0) + \frac{h^3}{3!} f'''(x_0) + \dots \dots \qquad (5)$$

2.1. Solution Procedure

1-Re-cast the given ordinary differential equation in the form given by equation (1).

2-Find at least three derivatives that appear in the expansion given by equation (5).

3-Evaluate all derived derivatives in step 3 at the point of expansion.

4-Make use of the given boundary condition, and the evaluated derivatives in Taylor's series.

Example

Solve the following boundary value problem $y' = -y$ at $x = 0.2$ given that $y(0) = 1$

Solution

The given O.D.E. is already of the required form, *i.e*, $y' = f(x, y)$

Let us follow the solution procedure described before, and start finding sufficient number of derivatives, leading to:

$$y' = -y$$

$$y'' = -y' = y$$

$$y''' = -y'' = y$$

$$\dots \dots \dots \dots \dots \dots$$

$$y^{(n)} = -y^{(n-1)}$$

Evaluate all the above derivatives at the point of expansion, as follows:

$$y(0) = 1$$

$$y'(0) = -y(0) = -1$$

$$y''(0) = 1$$

$$y'''(0) = -1$$

It is important to remember the fact that the step size has a direct effect on the accuracy of the solution, therefore due to this fact, let's get the solution one time with size step equal to 0.2 and once more with step size equal to 0.1 to see how decreasing size step affects improving the results.

Case (1) h = 0.2

Considering the point of expansion zero, we have:

$$h = x - x_0$$
$$= 0.2 - 0.0$$
$$= 0.2$$

$$f(x) = f(x_0) + \frac{h}{1!} f'(x_0) + \frac{h^2}{2!} f''(x_0) + \frac{h^3}{3!} f'''(x_0) + \dots$$

$$f(0.2) = 1 + \frac{(0.2)}{1!}(-1) + \frac{(0.2)^2}{2!}(1) + \frac{(0.2)^3}{3!}(-1)\dots = 0.81867$$

Case (2) h = 0.1

In this case, to find the required value, we should apply Taylor's series twice.

Consider the point of expansion zero, we have,

$$h = x - x_0 = 0.1 - 0.0 = 0.1$$

$$f(0.1) = 1 + \frac{(0.1)}{1!}(-1) + \frac{(0.1)^2}{2!}(1) + \frac{(0.1)^3}{3!}(-1)\ldots\ldots = 0.90481$$

The next step is to consider the point of expansion to be 0.1, therefore it is necessary to evaluate all derivatives at this new value, *i.e,*

$$y(0.1) = 0.90481$$

$$y'(0.1) = -0.90481$$

$$y''(0.1) = 0.90481$$

$$y'''(0.1) = -0/90481$$

Again, substituting into Taylor's series gives:

$$f(0.2) = 0.81873$$

The following tables compare the numerical results with the exact solution for this problem, see Table (**1**).

Table 1. Comparison between exact and numerical solution.

Differential equation	Exact Solution	Case (1) h = 0.2	Case (2) h = 0.1
$y' = -y$	$y = e^{-x}$ $\Rightarrow$ $y = e^{-0.2} = 0.81873$	0.81867	0.81873

It is clear from the table that when the step becomes small, the numerical solution becomes nearly the same as the exact solution. This is the first way of improving the accuracy of the solution by decreasing the step.

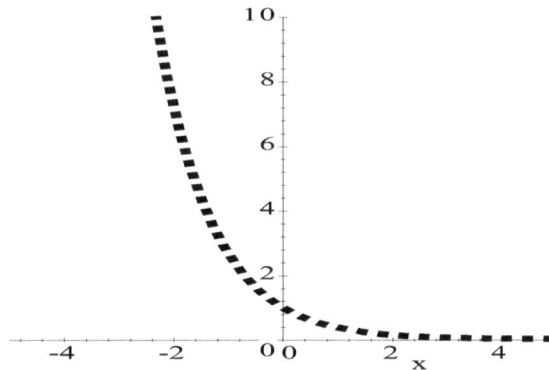

Fig. (1). Analytical solution.

3.TAYLOR'S SERIES METHOD FOR HIGHER ORDER O.D.E.

The same solution procedure described when using Taylor's series for solving first order differential equation [56] is still the same for solving higher order ordinary differential equation. We will illustrate how to use Taylor's series for solving higher order ordinary differential equation *via* the next example.

Example

Solve the following boundary value problem $y'' = x - y^2 + 3$ at $x = 0.1$ given that $y(0) = 1$ and $y'(0) = -2$

Solution

Start the solution by evaluating some higher derivatives, as follows:

$$y'' = x - y^2 + 3 \qquad (1)$$

$$y''' = 1 - 2yy' \qquad (2)$$

$$y^{(4)} = -2\left(yy'' + y'^2\right) \qquad (3)$$

Next, evaluating the above derivatives at the point of the expansion leads to:

$$y(0) = 1 \qquad \text{condition} \qquad (4)$$

$$y'(0) = -2 \quad \text{condition} \tag{5}$$

$$y''(0) = 2 \tag{6}$$

$$y'''(0) = 5 \tag{7}$$

$$y^{(4)}(0) = -12 \tag{8}$$

Take

$$\begin{aligned} h &= x - x_0 \\ &= 0.1 - 0.0 \\ &= 0.1 \end{aligned} \tag{9}$$

Substituting equations (4-8) into Taylor's expansion given by equation (5):

$$f(x) = f(x_0) + \frac{h}{1!} f'(x_0) + \frac{h^2}{2!} f''(x_0) + \frac{h^3}{3!} f'''(x_0) + \ldots \ldots \tag{10}$$

$$f(0.1) = 1 + \frac{(0.1)}{1!}(-2) + \frac{(0.1)^2}{2!}(2) + \frac{(0.1)^3}{3!}(5) + \frac{(0.1)^4}{4!}(-12)\ldots \tag{11}$$

$$= 0.810785$$

4. TAYLOR'S SERIES METHOD FOR SYSTEM OF 1ST ORDER O.D.E.

To understand the concept of solution of a system of two 1st order ordinary differential equations [57], let us assume that we have the following system as follows:

$$\frac{dx}{dt} = f_1(x, y, t) \tag{1}$$

$$\frac{dy}{dt} = f_2(x, y, t) \tag{2}$$

It is important to remember that when it is required to solve such system, boundary condition corresponding to each equation, should be provided separately.

Therefore, the associated two boundary conditions are as follows:

$$x(t = t_0) = x_0 \tag{3}$$

And

$$y(t = t_0) = y_0 \tag{4}$$

The basic idea of solution is to write Taylor's expansion for each equation, respectively taking a very important note into consideration that to what level of accuracy we will stop writing the expansion.

For example, if one wants to find the solution at the first level, then he should stop writing the expansion at the first derivative, and if the solution up to the second level is required, then he should stop at the second derivative and so on.

Example

Solve the following system of first order differential equations given as:

$$\frac{dx}{dt} = xy + t \tag{1}$$

$$\frac{dy}{dt} = x + yt \tag{2}$$

At $x = 0.1$ given that $x(0) = 1$ and $y(0) = -1$ (3)

Solution

First level of approximations:

$$x(t) = x(t_0) + \frac{h}{1!} x'(t_0) \tag{1}$$

$$y(t) = y(t_0) + \frac{h}{1!} y'(t_0) \tag{2}$$

And

$$h = t - t_0$$
$$= 0.1 - 0.0 \qquad (3)$$
$$= 0.1$$

$$\left(\frac{dx}{dt}\right)_{t=0} = (xy + t)_{t=0} = -1 \qquad (4)$$

$$\left(\frac{dy}{dt}\right)_{t=0} = (x + yt)_{t=0} = 1 \qquad (5)$$

Therefore,

$$x(0.1) = 1 + \frac{(0.1)}{1!}(-1) = 0.9 \qquad (6)$$

$$y(0.1) = -1 + \frac{(0.1)}{1!}(1) = -0.9 \qquad (7)$$

Second level of approximations:

Taylor's expansion takes the following form:

$$x(t) = x(t_0) + \frac{h}{1!}x'(t_0) + \frac{h^2}{2!}x''(t_0) \qquad (8)$$

$$y(t) = y(t_0) + \frac{h}{1!}y'(t_0) + \frac{h^2}{2!}y''(t_0) \qquad (9)$$

And

$$h = t - t_0$$
$$= 0.1 - 0.0 \qquad (10)$$
$$= 0.1$$

Evaluation the derivatives at the point of expansion:

$$\left(\frac{d^2x}{dt^2}\right)_{t=0} = \left(x\frac{dy}{dt} + y\frac{dx}{dt} + 1\right)_{t=0} = 3 \qquad (11)$$

$$\left(\frac{d^2y}{dt^2}\right)_{t=0} = \left(t\frac{dy}{dt} + y + \frac{dx}{dt}\right)_{t=0} = -2 \qquad (12)$$

$$x(0.1) = 1 + \frac{(0.1)}{1!}(-1) + \frac{(0.1)^2}{2!}(3) = 0.915 \qquad (13)$$

$$y(0.1) = -1 + \frac{(0.1)}{1!}(1) + \frac{(0.1)^2}{2!}(-2) = -0.910 \qquad (14)$$

Table 2. Summary of the results.

	First level of approximation		Second level of approximation	
$t = 0.1$	x	y	x	y
	0.9	-0.9	0.915	-0.910

5. EULER METHOD FOR 1ST ORDER O.D.E

We know now what does it mean by numerical solution of ordinary differential equation, we simply mean that we obtain the solution at specified points determined in advance.

Continuing the numerical methods for solving ordinary differential equations by introducing the Euler's method [80-82] and we will start by solving the first order differential equation.

Assume that we solved a first order differential equation, and the solution obtained the following points:

$$x_0 < x_1 < x_2 < \dots < x_N \le b \qquad (1)$$

Assume that:

$$x_n = x_o + nh, \quad n = 0,1,2,3,\ldots\ldots \tag{2}$$

Therefore, the solution will be of the form:

$$y(x_n) = y_h(x_n) = y_n \tag{3}$$

Now, let us derive Euler's method and let us assume that the differential equation will take the following form:

$$\frac{dY(x)}{dx} = f(x, Y(x)), \qquad x_0 \le x \le b \tag{4}$$

With the following associated boundary condition:

$$Y(x_0) = Y_0 \tag{5}$$

Assume that $Y(x)$ is the correct solution for the given first order differential equation.

Therefore, equation (4) can be re-cast in the following manner:

$$\frac{dY(x)}{dx} \cong \frac{1}{h}\left[Y(x+h) - Y(x)\right] \tag{6}$$

Note that at the point $x = x_n$, one can approximate the first derivative $\dfrac{dY(x)}{dx}$ as follows:

$$\frac{dY(x)}{dx} \cong f(x_n, Y(x_n)) \tag{7}$$

By making use of equation (7) into equation (6), the later will take the following form:

$$\frac{1}{h}\left[Y(x_{n+1}) - Y(x_n)\right] \cong f(x_n, Y(x_n)) \tag{8}$$

Simplification will lead to the following simplified form:

$$y_{n+1} = y_n + hf(x_n, y_n) \tag{9}$$

Equation (9) is the ***general Euler solution formula*** for the solution of first order differential equation.

Example

Given

$$y' = f(x, y) = x^2 + y \tag{1}$$

With

$$y(0) = 1 \tag{2}$$

Obtain y at $x = 0.02, 0.04, 0.06$ and 0.08

Solution

We have a first order differential equation, and it is required to find the solution at the indicated points shown in Fig. (**2**).

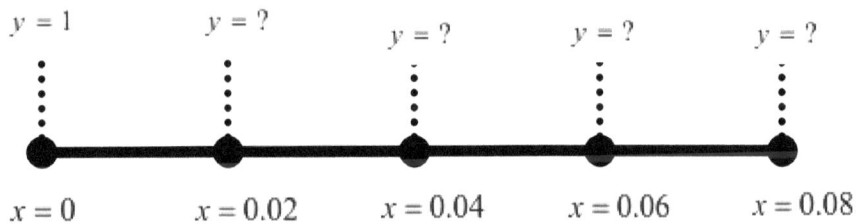

Fig. (2). Points of interest.

The first step of the solution is to rewrite the general Euler formula as follows:

$$y_{n+1} = y_n + h f(x_n, y_n) \tag{3}$$

With step size,

$$h = x_{n+1} - x_n \tag{4}$$

Put $n = 0$, this will lead to the solution at the point $x = 0.02$, therefore, the first step size will be:

$$
\begin{aligned}
h &= x_1 - x_0 \\
&= 0.02 - 0.0 \\
&= 0.02
\end{aligned}
\tag{5}
$$

Then the solution at that point denoted by $y(0.02)$ can be evaluated as follows:

$$
\begin{aligned}
y_1 &= y_0 + hf(x_0, y_0) = 1.0 + 0.02 \times \left(0^2 + 1\right) \\
&= 1.02
\end{aligned}
\tag{6}
$$

Repeat the previous step but now is for the next point, *i.e.*, put $n = 1$, this will lead to the solution at the point $x = 0.04$, therefore the second step size will be:

$$
\begin{aligned}
h &= x_2 - x_1 \\
&= 0.04 - 0.02 \\
&= 0.02
\end{aligned}
\tag{7}
$$

Then the solution at that point denoted by $y(0.04)$ can be evaluated as follows:

$$
\begin{aligned}
y_2 &= y_1 + hf(x_1, y_1) \\
&= 1.02 + 0.02 \times \left((0.02)^2 + 1.02\right) \\
&= 1.04041
\end{aligned}
\tag{8}
$$

Now repeat the same procedure up to the last point. These steps are summarized in Table (**3**) in the following manner.

6. EULER METHOD FOR HIGHER ORDER O.D.E.

The basic idea to apply Euler's method to solve higher order differential equation [83] is that, try as much as possible to reduce the order of the given ordinary differential equation to the first order, and then the same procedure for solving first order differential equation will be applied.

Let us try to show how to do so by solving the following example.

Table 3. Solution steps.

n	x_n	y_n
0	0.00	1.0000
1	0.02	1.0200
2	0.04	1.0404
3	0.06	1.0612
4	0.08	1.0825

Example

Given

$$y'' = f(x, y) = \frac{1}{2}\sqrt{x + y'} \qquad (1)$$

With

$$y'(0.40) = 0.41 \qquad (2)$$

Obtain

$\dfrac{dy}{dx}$ at $x = 0.6, 0.8$ and 1.0

Hint take $h = 0.1$

Solution

The first step of the solution as mentioned before is to reduce the order of the given differential equation, and to do so, let us assume the following:

$$u = y'$$
$$\Rightarrow \qquad (3)$$
$$u' = y''$$

By making use of equation (3), the given differential equation leads to:

$$u' = \frac{1}{2}\sqrt{x+u} \qquad (4)$$

$$u(0.40) = 0.41 \qquad (5)$$

Now let's write down the Euler's formula:

$$u_{n+1} = u_n + hf(x_n, u_n) = u_n + h\sqrt{(x_n + u_n)} \qquad (6)$$

Now we are ready to start computation, and instead of computations step by step, one can construct the table of results and fill it as mentioned before, see the following Table (4).

Table 4. Solution method.

n	x_n	u_n	u_{n+1}
0	0.4	0.41000	0.45500
1	0.5	0.45500	0.50386
2	0.6	0.50386	0.55639
3	0.7	0.55639	0.61244
4	0.8	0.61244	0.67186
5	0.9	0.67186	0.73455
6	1.0	0.73455	0.80040

7. EULER'S METHOD FOR SYSTEM OF FIRST O.D.E.

In the present section, we will learn how to solve system of first order differential equations using Euler's method. It will be also easy for the reader to do it *via* an example directly, specially, solving single equation like a system with no apparent difference stated.

Example

Solve the following system of first order differential equations given as:

$$\frac{dx}{dt} = f_1(x, y, t) \tag{1}$$

$$= x + y^2 + 3t$$

$$\frac{dy}{dt} = f_2(x, y, t) \tag{2}$$

$$= x^2 + y - t$$

$$x(t = 0) = 2 \tag{3}$$

$$y(t = 0) = 1 \tag{4}$$

Obtain the values of *x* and *y* at t = 0.2

Solution

The first and the basic step is to determine the size step, which can be carried out using the following relation:

$$h = t_{n+1} - t_n \tag{5}$$

For *n* = 0, equation (5) reduces to:

$$h = t_1 - t_0$$
$$= 0.2 - 0.0 \tag{6}$$
$$= 0.2$$

The next step is to write the Euler's general formula, taking into consideration that there is a dependent variable for each equation and one common independent variable, then:

$$x_{n+1} = x_n + hf_1(x_n, y_n, t_n) \tag{7}$$

$$y_{n+1} = y_n + hf_2(x_n, y_n, t_n) \tag{8}$$

The last step is the numerical computation, starting by $n = 0$ as follows:

Equations (7-8) will be:

$$x_1 = x_0 + hf_1(x_0, y_0, t_0) = x_0 + h(x_0 + y_0^2 + 3t_0) \tag{9}$$

$$y_1 = y_0 + hf_2(x_0, y_0, t_0)$$
$$= y_0 + h(x_0^2 + y_0 - t_0)$$ (10)

Computations lead to:

$$x_1 = 2.60$$ (11)

$$y_1 = 2.00$$ (12)

Repeat the same procedure till we reach the end point of computation.

8. MODIFIED EULER'S METHOD FOR FIRST O.D.E.

It is clear from the above methods that no modification occurs for the resulted values, and so there will be some error. The modified Euler [84] is a method in which successive approximation will occur for the computed result up to a prescribed accuracy is obtained.

8.1. Solution Procedure

1-Start with the usual formula for Euler to find the first approximation for the solution.

2-Evaluate the average for the 1st derivative and the derivative from the first approximation obtained in step (1).

3-Use the average evaluated from step (2) to evaluate the approximation for the 2nd time.

4-Repeat the steps (1-3) till we obtain the pre-scribed accuracy or up to a prescribed number of iteration.

The next example will illustrate the solution procedure.

Example

Given:

$$y' = f(x, y)$$
$$= x^2 + y$$ (1)

With

$$y(0) = 1 \tag{2}$$

Use modified Euler's formula to obtain y at $x = 0.1$ up to the third approximation.

Solution

Let us start the solution by re-writing the normal Euler's formula but we will make some new modification, *i.e.*, instead of writing y_1 we will write it as $y_1^{(1)}$ to get successive approximation, therefore:

$$y_1^{(1)} = y_0 + hf(x_0, y_0) \tag{3}$$

And

$$\begin{aligned} h &= x_1 - x_0 \\ &= 0.1 - 0.0 \\ &= 0.1 \end{aligned} \tag{4}$$

Therefore, the first approximation will be:

$$\begin{aligned} y_1^{(1)} &= 1.0 + (0.1)\{(0.0)^2 + 1.0\} \\ &= 1.10 \end{aligned} \tag{5}$$

The next step is to write the general modified Euler's method, as follows:

$$y_i^{(n+1)} = y_{i-1} + \frac{h}{2}\left[f(x_{i-1}, y_{i-1}) + f((x_{i-1} + h), y_i^{(n)})\right] \tag{6}$$

It is important to say some notes appeared in equation (6), there are two indices the first is n which represents the number of iterations and the second is i which is used to differentiate the successive values of the dependent variable. For example, if one uses $n = 1$ & $i = 1$, then we can write equation (6) in the following manner:

$$y_1^{(2)} = y_0 + \frac{h}{2}\left[f(x_0, y_0) + f((x_0 + h), y_1^{(1)})\right] \tag{7}$$

Equation (7) $y_1^{(1)}$ is the first approximation for the first value of the dependent variable y, while $y_1^{(2)}$ is the second approximation for the first value of the dependent variable y, and so on. Numerically the second approximation will take the following form:

$$y_1^{(2)} = 1.0 + \left(\frac{0.1}{2}\right)\left[\left(0^2 + 1.0\right) + \left\{(0.0 + 0.1)^2 + 1.1\right\}\right]$$

$$= 1.1055$$

(8)

Put $n = 2 \ \& \ i = 1$, one can get the third approximation of the dependent variable at the same point number one.

$$y_1^{(3)} = y_0 + \frac{h}{2}\left[f(x_0, y_0) + f\left((x_0 + h), y_1^{(2)}\right)\right]$$

(9)

$$y_1^{(3)} = 1.0 + \left(\frac{0.1}{2}\right)\left[\left(0^2 + 1.0\right) + \left\{(0.0 + 0.1)^2 + 1.1055\right\}\right]$$

$$= 1.10578$$

(10)

The above computations are summarized in the next Fig. **(3)**, so as it can be an excellent guide for the reader.

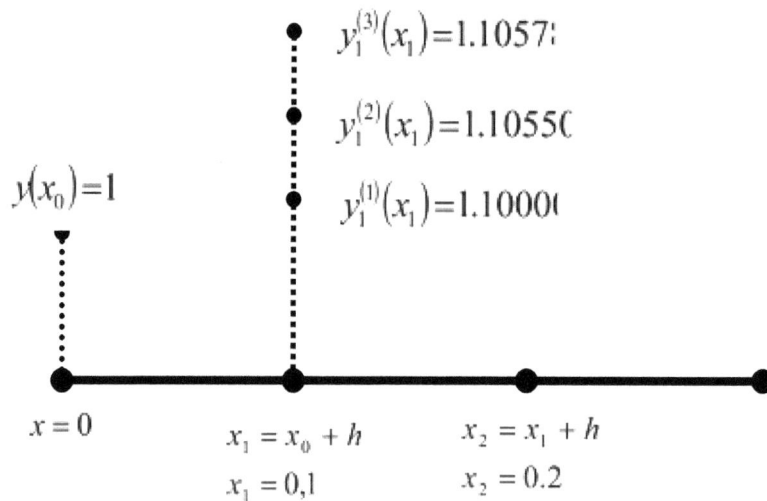

Fig. (3). Modified Euler's method.

9. MODIFIED EULER'S METHOD FOR HIGHER ORDER O.D.E.

The basic idea is to reduce the order of the differential equation to make it first order, and then the same procedure will be applied.

The next example is designated to illustrate the solution procedure.

Example

Given

$$y'' = f(x,y) = \frac{1}{2}\sqrt{x + y'} \qquad (1)$$

With

$$y'(0.40) = 0.41 \qquad (2)$$

Obtain $\dfrac{dy}{dx}$ at $x = 0.6, 0.8$ and 1.0 Hint take $h = 0.1$

Solution

Assume that:

$$u = y'$$
$$\Rightarrow \qquad (3)$$
$$u' = y''$$

Re-casting the given differential equation leads to:

$$u' = \frac{1}{2}\sqrt{x + u} \qquad (4)$$

$$u(0.40) = 0.41 \qquad (5)$$

Therefore, the given differential equation takes the following new form in terms of the new variable. *The remaining steps for solution are left as an exercise.*

10. MODIFIED EULER'S METHOD FOR SOLVING SYSTEM OF FIRST ORDER O.D.E.

In the present section, we will apply the modified Euler's method to solve a system of two ordinary first order differential equations of the following form with their associated boundary conditions:

$$\frac{dx}{dt} = f_1(x, y, t) \tag{1}$$

$$\frac{dy}{dt} = f_2(x, y, t) \tag{2}$$

With

$$x(t = 0) = x_0 \tag{3}$$

$$y(t = 0) = y_0 \tag{4}$$

The general formula for approximating the solution up to $(n+1)$ order will take the following form:

$$x_i^{(n+1)} = x_{i-1} + \frac{h}{2}\left[f_1\left(x_i^{(n)}, y_i^{(n)}, t_{n+1}\right) + f_2\left(x_i^{(n)}, y_i^{(n)}, t_{n+1}\right)\right] \tag{5}$$

$$y_i^{(n+1)} = y_{i-1} + \frac{h}{2}\left[f_1\left(x_i^{(n)}, y_i^{(n)}, t_{n+1}\right) + f_2\left(x_i^{(n)}, y_i^{(n)}, t_{n+1}\right)\right] \tag{6}$$

In these equations

$$h = t_{n+1} - t_n \tag{7}$$

Example

Solve the following system of first order differential equations given as:

$$\frac{dy}{dx} = f_1(x, y, z) \tag{1}$$

$$= y - z + 6x$$

$$\frac{dz}{dx} = f_2(x, y, z) \tag{2}$$

$$= y^2 + z - 5x$$

$$y(x = 0) = 3 \tag{3}$$

$$z(x = 0) = 4 \tag{4}$$

Obtain the values of y and z at $x = 0.2$

Solution

Start the solution by writing the first order approximation for both the two dependent variables as follows:

$$y_{n+1}^1 = y_n + h(y_n - z_n + 6x_n) \tag{5}$$

$$z_{n+1}^1 = z_n + h\left(y_n^2 + z_n - 5x_n\right) \tag{6}$$

Put $n = 0$

$$y(x = 0) = y_0 = 3 \tag{7}$$

$$z(x = 0) = z_0 = 4 \tag{8}$$

The next step is to determine the step size, which is the difference between two successive values of the independent variable, as follows:

$$x_1 - x_0 = 0.2 - 0.0 = 0.2 \tag{10}$$

Now:

$$y_1^1 = y_0 + h(y_0 - z_0 + 6x_0) \Rightarrow y_1^1 = 2.8 \tag{11}$$

$$z_1^1 = z_0 + h\left(y_0^2 + z_0 - 5x_0\right) = z_1^1 = 6.6 \tag{12}$$

Now the approximation of order *m* is as follows:

$$y_{n+1}^m = y_n^{m-1} + \left(\frac{h}{2}\right)\left[\left(y_n^{m-1} - z_n^{m-1} + 6x_n\right) + \left(y_n - z_n + 6x_n\right)\right] \tag{13}$$

$$z_{n+1}^m = z_n + \frac{h}{2}\left[\left(y_n^{2(m-1)} + z_n^{(m-1)} - 5x_n\right) + \left(y_n^2 + z_n - 5x_n\right)\right] \tag{14}$$

Summary of the computed results are shown in the next Table (**6**).

Table 6. Computed results.

$y_1^1 = 2.80000$	$z_1^1 = 6.60000$
$y_1^2 = 2.64000$	$z_1^2 = 6.644000$
$y_1^3 = 2.6196$	$z_1^3 = 6.56140$
$y_1^4 = 2.62582$	$z_1^4 = 6.54237$
$y_1^5 = 2.62835$	$z_1^5 = 6.54373$
$y_1^6 = 2.62833$	$z_1^6 = 6.54519$

11. RUNGE-KUTTA METHOD FOR FIRST O.D.E.

To understand the Runge-Kutta method [85, 86], let us explain the solution procedure applied to a first order ordinary differential equation as follows:

$$y' = f(x, y) \tag{1}$$

With boundary condition of the form:

$$y(x_0) = y_0 \tag{2}$$

11.1. Solution Procedure

The required solution at the point $x_1 = x_0 + h$ can be evaluated from the following equation:

$$y_1 = y_0 + K_{ave} \tag{3}$$

Where

$$K_{ave} = \frac{1}{6}\{K_1 + 2 \bullet (K_2 + K_3) + K_4\} \tag{4}$$

And

$$K_1 = hf(x_0, y_0) \tag{5}$$

$$K_2 = hf\left(x_0 + \frac{h}{2}, y_0 + \frac{K_1}{2}\right) \tag{6}$$

$$K_3 = hf\left(x_0 + \frac{h}{2}, y_0 + \frac{K_2}{2}\right) \tag{7}$$

$$K_4 = hf(x_0 + h, y_0 + K_3) \tag{8}$$

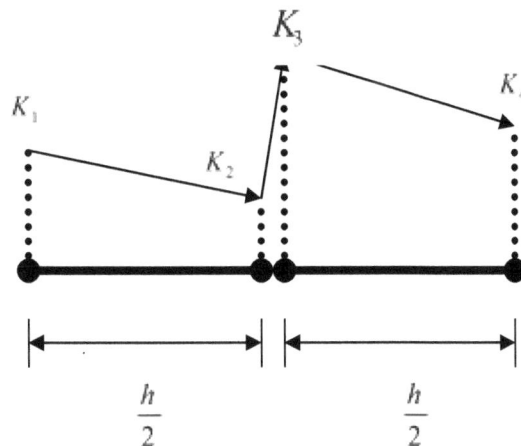

Fig. (4). Runge-Kutta method.

Fig. (**4**) shows the procedure for evaluating the successive values of the variables $K_i, i = 1,2,3,4$ and their locations.

Example

Given:

$$\frac{dy}{dx} = f(x,y) = y^2 + 3x \tag{1}$$

$$y(x_0 = 0) = y_0 = 1 \tag{2}$$

Use Runge-Kutta method to obtain the values of y at $x = 0.1$

Solution

As usual the first step is to determine the step size, as follows:

$$\begin{aligned} h &= x_1 - x_0 \\ &= 0.1 - 0.0 \\ &= 0.1 \end{aligned} \tag{3}$$

The next step is to evaluate the different values of the variables $K_i, i = 1,2,3,4$

$$K_1 = hf(x_0, y_0) \tag{4}$$

$$K_1 = 0.1\left[(0)^2 + (1)^2\right] = 0.1 \tag{5}$$

$$K_2 = hf\left(x_0 + \frac{h}{2}, y_0 + \frac{K_1}{2}\right) \tag{6}$$

$$K_2 = (0.1) \times \left[3 \times \left(0 + \frac{0.1}{2}\right) + \left(1 + \frac{0.1}{2}\right)^2\right] \tag{7}$$

$$= 0.12525$$

$$K_3 = hf\left(x_0 + \frac{h}{2}, y_0 + \frac{K_2}{2}\right) \tag{8}$$

$$K_3 = (0.1) \times \left[3 \times \left(0 + \frac{0.1}{2} \right) + \left(1 + \frac{0.12525}{2} \right)^2 \right] \tag{9}$$

$$= 0.12791718$$

$$K_4 = hf(x_0 + h, y_0 + K_3) \tag{10}$$

$$K_4 = (0.1) \times \left[3 \times (0 + 0.1) + (1 + 0.12791718)^2 \right] \tag{11}$$

$$= 0.15721969$$

Now

$$K_{ave} = \frac{1}{6} \{ K_1 + 2 \times (K_2 + K_3) + K_4 \} \tag{12}$$

$$K_{ave} = 0.12725896 \tag{13}$$

Therefore

$$y_1 = y_0 + K_{ave}$$
$$\Rightarrow \tag{14}$$
$$y_1 = 1.12725896$$

12. RUNGE-KUTTA METHOD FOR SYSTEM OF FIRST O.D.E.

The same procedure for solving single first order ordinary differential equation can be applied when solving system of two first order ordinary differential equations. Let us illustrate the procedure on the next example.

$$\frac{dx}{dt} = f_1(x, y, t) \tag{1}$$

$$\frac{dy}{dt} = f_2(x, y, t) \tag{2}$$

With

$$x(t = 0) = x_0 \tag{3}$$

$$y(t = 0) = y_0 \tag{4}$$

The solution for both x and y based on Runge-Kutta method, without proof takes the following form:

$$x_1 = x_0 + \left(K_{ave}\right)_x \tag{5}$$

And

$$y_1 = y_0 + \left(K_{ave}\right)_y \tag{6}$$

Where

$$\left(K_{ave}\right)_x = \frac{1}{6}\left[\left(K_1\right)_x + 2 \times \left[\left(K_2\right)_x + \left(K_3\right)_x\right] + \left(K_4\right)_x\right] \tag{7}$$

$$\left(K_{ave}\right)_y = \frac{1}{6}\left[\left(K_1\right)_y + 2 \times \left[\left(K_2\right)_y + \left(K_3\right)_y\right] + \left(K_4\right)_y\right] \tag{8}$$

$$\left(K_1\right)_x = hf_1\left(t_0, x_0, y_0\right) \tag{9}$$

$$\left(K_1\right)_y = hf_2\left(t_0, x_0, y_0\right) \tag{10}$$

$$\left(K_2\right)_x = hf_1\left(\left[t_0 + \frac{h}{2}\right], \left[x_0 + \frac{\left(K_1\right)_x}{2}\right], \left[y_0 + \frac{\left(K_1\right)_y}{2}\right]\right) \tag{11}$$

$$\left(K_2\right)_y = hf_2\left(\left[t_0 + \frac{h}{2}\right], \left[x_0 + \frac{\left(K_1\right)_x}{2}\right], \left[y_0 + \frac{\left(K_1\right)_y}{2}\right]\right) \tag{12}$$

$$\left(K_3\right)_x = hf_1\left(\left[t_0 + \frac{h}{2}\right], \left[x_0 + \frac{\left(K_2\right)_x}{2}\right], \left[y_0 + \frac{\left(K_2\right)_y}{2}\right]\right) \tag{13}$$

$$\left(K_3\right)_y = hf_2\left(\left[t_0 + \frac{h}{2}\right], \left[x_0 + \frac{\left(K_2\right)_x}{2}\right], \left[y_0 + \frac{\left(K_2\right)_y}{2}\right]\right) \tag{14}$$

$$(K_4)_x = hf_1\left(\left[t_0 + h\right], \left[x_0 + (K_3)_x\right], \left[y_0 + (K_3)_y\right]\right) \tag{15}$$

$$(K_4)_y = hf_2\left(\left[t_0 + h\right], \left[x_0 + (K_3)_x\right], \left[y_0 + (K_3)_y\right]\right) \tag{16}$$

Example

Solve the following system of first order differential equations given as:

$$\frac{dx}{dt} = f_1(x, y, t) \tag{1}$$
$$= x^2 + y + 5t$$

$$\frac{dy}{dt} = f_2(x, y, t) \tag{2}$$
$$= x + y^2 - 6t$$

$$x(t = 0) = 2 \tag{3}$$

$$y(t = 0) = 3 \tag{4}$$

Obtain the values of x and y at t = 0.2

Solution

Follow up the procedure described before as:

$$x_1 = x_0 + (K_{ave})_x \tag{5}$$

$$y_1 = y_0 + (K_{ave})_y \tag{6}$$

$$(K_{ave})_x = \frac{1}{6}\left[(K_1)_x + 2 \times \left[(K_2)_x + (K_3)_x\right] + (K_4)_x\right] \tag{7}$$

$$(K_{ave})_y = \frac{1}{6}\left[(K_1)_y + 2 \times \left[(K_2)_y + (K_3)_y\right] + (K_4)_y\right] \tag{8}$$

$$(K_1)_x = hf_1(t_0, x_0, y_0) \tag{9}$$

$$\left(K_1\right)_x = (0.2) \times \left[2^2 + 3 + 0\right] = 1.4 \tag{10}$$

$$\left(K_1\right)_y = hf_2\left(t_0, x_0, y_0\right) \tag{11}$$

$$\left(K_1\right)_y = (0.2) \times \left[2 + 3^2 - 0\right] = 2.2 \tag{12}$$

$$\left(K_2\right)_x = hf_1\left(\left[t_0 + \frac{h}{2}\right], \left[x_0 + \frac{\left(K_1\right)_x}{2}\right], \left[y_0 + \frac{\left(K_1\right)_y}{2}\right]\right) \tag{13}$$

$$\left(K_2\right)_x = (0.2) \times \left(5\left[0 + \frac{0.2}{2}\right] + \left|2 + \left(\frac{1.4}{2}\right)^2\right| + \left[3 + \frac{2.2}{2}\right]\right) \tag{14}$$
$$= 2.378$$

$$\left(K_2\right)_y = hf_2\left(\left[t_0 + \frac{h}{2}\right], \left[x_0 + \frac{\left(K_1\right)_x}{2}\right], \left[y_0 + \frac{\left(K_1\right)_y}{2}\right]\right) \tag{15}$$

$$\left(K_2\right)_y = (0.2) \times \left(-6\left[0 + \frac{0.2}{2}\right] + \left[2 + \frac{1.4}{2}\right] + \left|3 + \left(\frac{2.2}{2}\right)^2\right|\right) \tag{16}$$
$$= 3.782$$

$$\left(K_3\right)_x = hf_1\left(\left[t_0 + \frac{h}{2}\right], \left[x_0 + \frac{\left(K_2\right)_x}{2}\right], \left[y_0 + \frac{\left(K_2\right)_y}{2}\right]\right) \tag{17}$$

Similarly one can compute other yields as follows:

$$\left(K_3\right)_x = 3.11214 \tag{18}$$

$$\left(K_3\right)_y = 5.3028 \tag{19}$$

$$\left(K_4\right)_x = 7.10629 \tag{20}$$

$$\left(K_4\right)_y = 14.56765 \tag{21}$$

Now

$$\left(K_{ave}\right)_x = 3.24776 \tag{22}$$

$$\left(K_{ave}\right)_y = 5.822668 \tag{23}$$

Therefore

$$x_1 = 5.247758 \tag{24}$$

$$y_1 = 8.822668 \tag{25}$$

13. APPLICATIONS OF LINEAR DIFFERENTIAL EQUATIONS

The linear differential equations with constant coefficients find their most important applications in the study of electrical, mechanical and other linear systems.

In the present chapter, we shall study some mechanical and electrical engineering applications, starting from explaining the types of oscillations of mechanical systems and the equivalent electrical circuits.

13.1. Simple Harmonic Motion

When the acceleration of a particle is proportional to its displacement from a fixed point and directed towards it, then the motion is called simple *harmonic motion* [87].

If the displacement of the particle at any time t, from the fixed point O is x, see the next Fig. (5):

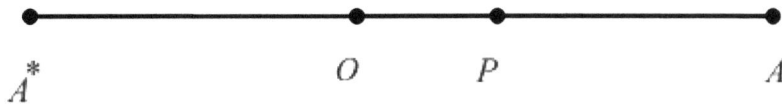

Fig. (5). Harmonic motion.

Then:

$$\frac{d^2 x}{dt^2} = -\mu^2 x$$

Its solution will be:

$$x = c_1 \cos \mu t + c_2 \sin \mu t$$

Its velocity at P will be:

$$\frac{dx}{dt} = \mu(-c_1 \sin \mu t + c_2 \cos \mu t)$$

Example

In case of a stretched elastic horizontal string, which has one end fixed and a particle of mass m attached to the other end,

prove that the equation of motion is $\dfrac{d^2 s}{dt^2} = -g(s - \ell)$ where, ℓ is the natural length of the string and e is the elongation due to weight.

Solution

Let $OA = \ell$ be the elastic horizontal string with end O fixed and having a particle of mass m attached to the end A. At any time t, let the particle be at P, $OP = s$ so that the elongation $AP = s - \ell$.

Since for elongation e, tension is equal to mg, then for elongation $s - \ell$, tension will be equal to $\dfrac{mg(s - \ell)}{e}$.

Tension being the only horizontal force, the equation of the motion takes the following form:

$$m\frac{d^2 s}{dt^2} = -T$$

$$\frac{d^2 s}{dt^2} = -\frac{T}{m}$$

$$= -\frac{g(s - \ell)}{e}$$

$$e\frac{d^2 s}{dt^2} = -g(s - \ell)$$

This is the required equation of motion. The complete solution is given as:

$$s = c_1 \cos\sqrt{\left(\frac{g}{e}\right)}t + c_2 \sin\sqrt{\left(\frac{g}{e}\right)}t + \ell$$

To find the constants in the obtained solution, one can apply the initial conditions, but in our present example, it is enough to obtain the solution in a general form.

13.2. Oscillation of a Spring

13.2.1. Free Oscillation

Suppose that a mass m is suspended from the end A of a light spring the other end of which is fixed at O. Let $e = AB$ be the elongation produced by the mass m hanging in equilibrium.

If k be the restoring force per unit stretch of the spring due to elasticity, then for the equilibrium at B takes the form:

$$mg = T = ke$$

At any time t, after the motion ensues, let the mass be at P where $BP = x$. Then the equation of motion is given as:

$$m\frac{d^2x}{dt^2} = mg - k(e + x) = -kx$$

If one assumes

$$\frac{k}{m} = \mu^2$$

Then

$$\frac{d^2x}{dt^2} + \mu^2 x = 0$$

This equation represents the free vibrations of the spring, which are of the simple harmonic form having center of oscillation at B and period of oscillation of the form given as:

Period of oscillatio n $= \dfrac{2\pi}{\mu} = 2\pi\sqrt{\left(\dfrac{e}{g}\right)}$

13.2.2. Damped Oscillation

If the motion of the mass m is subject to an additional force resistance, the oscillations are said to be **damped**. The damping force may be constant or proportional to velocity. The latter type of damping is important and usually called **viscous damping**.

Now if the damping force be proportional to velocity, say, $= r\dfrac{dx}{dt}$ then the equation of motion becomes:

$$m\frac{d^2x}{dt^2} = mg - k(e+x) - r\frac{dx}{dt}$$

$$= -kx - r\frac{dx}{dt}$$

Or writing $\dfrac{r}{m} = 2\lambda$ and $\dfrac{k}{m} = \mu^2$

Then, the equation of motion becomes:

$$\frac{d^2x}{dt^2} + 2\lambda\frac{dx}{dt} + \mu^2 x = 0$$

Its auxiliary equation will be:

$$D^2 + 2\lambda D + \mu^2 = 0$$

Then

$$D = -\lambda \pm \sqrt{\lambda^2 - \mu^2}$$

Three different cases appear when analyzing the above equation.

Case (1)

When $\lambda > \mu$, the roots of the auxiliary equation are real and distinct say γ_1 and γ_2. The solution of equation of motion takes the form:

$$x = c_1 e^{\gamma_1 t} + c_2 e^{\gamma_2 t}$$

To determine the constants c_1 and c_2, let the spring be stretched to a length ℓ and then released so that $x = \ell$ and $\dfrac{dx}{dt} = 0$ at $t = 0$.

Then:

$$\ell = c_1 + c_2$$

In addition, from $\dfrac{dx}{dt}$ we get:

$$c_1 \gamma_1 + c_2 \gamma_2 = 0$$

Then

$$c_1 = \frac{-\ell \gamma_2}{\gamma_1 - \gamma_2}$$

And

$$c_2 = \frac{-\ell \gamma_1}{\gamma_1 - \gamma_2}$$

The restoring force in this case, is so great that the motion is non-oscillatory and therefore, referred to as ***over-damped*** or ***dead-beat*** motion.

Case (2)

When $\lambda = \mu$, the roots of the auxiliary equation are real and equal each being $= -\lambda$, the solution of equation of motion takes the form:

$$x = (c_1 + c_2 t)e^{-\lambda t}$$

Following the same procedure given above to determine the constants c_1 and c_2, the complete solution in this case takes the following form:

$$x = \ell(1 + \lambda t)e^{-\lambda t}$$

The nature of the motion is similar to that of the previous case and is called the *critically damped* motion.

Case (3)

When $\lambda < \mu$, the roots of the auxiliary equation are imaginary and equal each being $D = -\lambda \pm i\alpha$, in which $\alpha = \mu^2 - \lambda^2$. The solution of equation of motion takes the form:

$$x = e^{-\lambda t}(c_1 \cos \alpha t + c_2 \sin \alpha t)$$

Following the same procedure given above to determine the constants c_1 and c_2, the complete solution in this case takes the following form:

$$x = \ell \sqrt{\left(1 + \frac{\lambda^2}{\alpha}\right)} e^{-\lambda t} \cos\left(\alpha t - \tan^{-1}\frac{\lambda}{\alpha}\right)$$

Note that, the presence of trigonometric factor shows that the motion is oscillatory.

13.2.3. Forced Oscillation Without Damping

If the point of the support is also vibrating with some external periodic force, then the resulting motion is called forced oscillatory motion. Taking the external periodic force to be $mp \cos nt$, then the equation of motion is given as:

$$m\frac{d^2 x}{dt^2} = mg - k(e + x) + mp \cos nt$$

But $mg = ke$ and writing $\dfrac{k}{m} = \mu^2$

Then, the equation of motion becomes:

$$\frac{d^2 x}{dt^2} + \mu^2 x = p \cos nt$$

The complementary solution will be:

$$x_{c.s.} = c_1 \cos \mu t + c_2 \sin \mu t$$

In addition, the particular solution will be:

$$x_{p.s.} = p\left(\frac{1}{D^2 + \mu^2}\right) \cos nt$$

Therefore, two different cases appear as follows:

Case (1) $\mu \neq n$

$$x_{p.s.} = p\left(\frac{1}{\mu^2 - n^2}\right) \cos nt$$

Then, the complete solution will be:

$$x_{G.S.} = c_1 \cos \mu t + c_2 \sin \mu t + \frac{p}{\mu^2 - n^2} \cos nt$$

This shows that two oscillatory motions compound the motion: the first due to complementary function, which gives free oscillation and the second due to particular solution and gives forced oscillation.

Case (2) $\mu = n$

$$x_{p.s.} = pt\left(\frac{1}{2D}\right) \cos \mu t = \frac{pt}{2} \int \cos \mu t \, dt = \frac{pt}{2\mu} \sin \mu t$$

Then, the complete solution is given as:

$$x_{G.S.} = c_1 \cos \mu t + c_2 \sin \mu t + \frac{pt}{2\mu} \sin \mu t$$

This shows that the oscillations have

Period $\dfrac{2\pi}{\mu}$

Amplitude $\rho = \sqrt{\left(c_2 + \dfrac{pt}{2\mu}\right)^2 + c_1^2}$

Thus, the amplitude of the oscillations may become abnormally large causing over-strain and consequently break down of the system.

This phenomenon of the impressed frequency becomes equal to the natural frequency of the system, referred to as **resonance**. Thus, while designing a machine or a structure, the occurrence of resonance should be avoided to check the rupture of the system at any stage.

13.2.4. Forced Oscillation with Damping

If in addition to the external periodic force to be $mp \cos nt$, there is a damping force proportional to the velocity, then the equation of motion will be:

$$m\frac{d^2x}{dt^2} = mg - k(e + x) + mp \cos nt - r\frac{dx}{dt}$$

But $mg = ke$ and writing $\dfrac{k}{m} = \mu^2$, and by writing $\dfrac{r}{m} = 2\lambda$

The equation of motion becomes:

$$\frac{d^2x}{dt^2} + 2\lambda\frac{dx}{dt} + \mu^2 x = p \cos nt$$

The complementary solution will be:

$$x_{c.s.} = e^{-\lambda t}\left(c_1 e^{t\sqrt{\left(\lambda^2 - \mu^2\right)}} + c_2 e^{-t\sqrt{\left(\lambda^2 - \mu^2\right)}} \right)$$

It represents the free oscillations of the system, which die out as the time tends to infinity. In addition, the particular solution is given as:

$$x_{p.s.} = p\left(\frac{1}{D^2 + 2\lambda D + \mu^2} \right)\cos nt$$

$$= p\frac{\left(\mu^2 - n^2\right)\cos nt + 2\lambda n \sin nt}{\left(\mu^2 - n^2\right)^2 + 4\lambda^2 n^2}$$

This represents the forced oscillations of the system having:

Period $\dfrac{2\pi}{n}$

Amplitude $\rho = \dfrac{p}{\sqrt{\left(\mu^2 - n^2\right)^2 + 4\lambda^2 n^2}}$

Example

A body weighting 3 lb is hung from a spring. A pull of 25 lb weight will stretch the spring to 3 inches below the static equilibrium position and then released. Find the displacement of the body from its equilibrium position at any time, the maximum velocity and the period of oscillation.

Solution

Configuration of the problem is shown in the following figure.

Let O be a fixed end and A be the lower end of the spring. Since a pull of 25 lb weight at the end A stretches the spring 3 inches which equals 3 inches $= \dfrac{2}{4} ft$.

Then:

$$25 = T_0 = k\left(\frac{1}{4} \right) \Rightarrow k = 100 \text{ lb/ft}$$

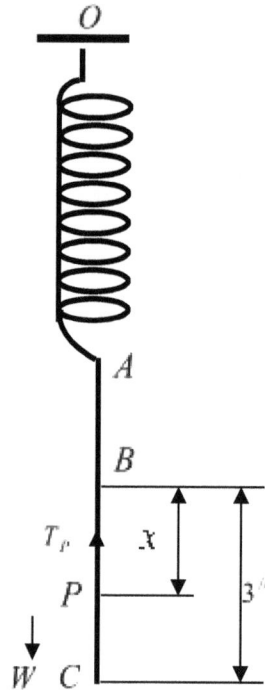

Fig. (6). Example configuration

Let B be the equilibrium position when a body weighting $8\,\text{lb}$ is hung from A, then:

$$8 = T_B = k(AB) \Rightarrow AB = \frac{8}{100} = 0.08 \text{ ft}$$

Now the weight is pulled down to C, where $BC = \dfrac{1}{4}$ ft. After a time t of its release from C, let the weight be at P, where $BP = x$, then the tension $T_P = k \times AP = 8 + 100x$.

Then the equation of the motion of the body is given as:

$$\frac{W}{g}\frac{d^2x}{dt^2} = W - T_P$$

$$\frac{8}{32}\frac{d^2x}{dt^2} = 8 - (8 + 100x)$$

$$\frac{d^2x}{dt^2} = \mu^2 x$$

This shows that the motion of the body is simple harmonic about B as a center and the period of oscillation will be $\dfrac{2\pi}{\mu} = \dfrac{\pi}{10}$.

In addition, the amplitude of the motion being $BC = \dfrac{1}{4}$ ft the displacement of the body from B at a time t is given by:

$$x = \frac{1}{4}\cos\mu t = \frac{1}{4}\cos(20)\,t$$

And the maximum velocity

$$\frac{dx}{dt} = \mu(\text{amplitude})) = 20 \times \frac{11}{44} = 5 \text{ ft/sec}$$

Example

A light elastic string of natural length ℓ has one end fixed at a point A and the other end attached to a stone the weight of which, in equilibrium, would extend the string to a length ℓ_1. Show that if the stone be dropped from rest at A it will come to an instantaneous rest at a depth $\sqrt{(\ell_1^2 - \ell^2)}$ below the equilibrium position.

Solution

Let $AB = \ell$ be the natural length of the string. Let AC be the extended string when the stone of the mass m hangs in equilibrium, so that $BC = \ell_1 - \ell$.

If λ be the modulus of elasticity, then for the equilibrium of C, we have:

$$mg = T = \frac{\lambda(\ell_1 - \ell)}{\ell}$$

When the motion ensues, let P be the position of the stone at any time, such that $CP = x$.

The equation of the motion will be:

$$m\frac{d^2x}{dt^2} = mg - T_1 = mg - \frac{\lambda(\ell_1 - \ell) - \ell}{\ell}$$

$$= mg - \frac{\lambda(\ell_1 - \ell)}{\ell} - \frac{\lambda x}{\ell} = -\left\{\frac{mg}{(\ell_1 - \ell)}\right\}x$$

Then

$$\frac{d^2x}{dt^2} = -\mu^2 x$$

Where

$$\mu^2 = \frac{g}{(\ell_1 - \ell)}$$

This shows that the motion is simple harmonic having center at C, the position of equilibrium of the stone. The stone falls freely under gravity up to B.

Then the velocity acquired at B is given as:

$$V_B = \sqrt{2g\ell}$$

From equations (i), (ii), the velocity at P is given by:

$$V^2 = \mu^2\left(a^2 - x^2\right)$$

Where a is the amplitude

At B

$$V_B = \sqrt{2g\ell}$$
$$x = CB = -(\ell_1 - \ell)$$

Then

$$2g\ell = \mu^2\left[a^2 - (\ell_1 - \ell)^2\right]$$
$$\Rightarrow$$
$$a = \sqrt{\left(\ell_1^2 - \ell^2\right)}$$

Hence, the stone comes to rest at a distance $\sqrt{\left(\ell_1^2 - \ell^2\right)}$ below the equilibrium position.

13.3. Electrical Applications

In this section, we consider simple electric circuits containing a resistor and an inductor or capacitor in series with a source of electromotive force (emf) [88].

•An electromotive force, E (volts), usually a battery or generator, drives an electric charge Q (coulombs) and produces a current I (amperes). The current is defined as the rate of flow of the charge, and we can write:

$$I = \frac{dQ}{dt}$$

•A resistor of resistance R (ohms) is a component of the circuit that opposes the current, dissipating the energy in the form of heat. It produces a drop in voltage given by Ohm's law:

$$E_R = RI$$

•An inductor of inductance L (henrys) opposes any change in the current by producing a voltage drop of:

$$E_L = L\frac{dI}{dt}$$

•A capacitor C (farads) stores charge. In doing so, it resists the flow of further charge, causing a drop in the voltage of:

$$E_C = \frac{Q}{C}$$

13.3.1. R. L. C Circuit

If we apply Kirchoff's law to the circuit above, we can obtain:

$$RI + \frac{1}{C}\int I\, dt + L\frac{dI}{dt} = E\ (t) \tag{a}$$

But

$$I = \frac{dQ}{dt}$$

In which $Q(t)$ is the charge. Then:

$$R\frac{dQ}{dt} + \frac{Q}{C} + L\frac{d^2Q}{dt^2} = E(t) \tag{b}$$

Or

$$\frac{d^2Q}{dt_2} + \frac{R}{L}\frac{dQ}{dt} + \frac{Q}{LC} = \frac{E(t)}{L}$$

This is a second order D. E. In most practical problems, not the charge $Q(t)$ but the current $I(t)$ is the physical quantity of interest.

Differentiating equation (a) one can obtain:

$$L\frac{d^2I}{dt^2} + R\frac{dI}{dt} + \frac{1}{C}I = \frac{dE(t)}{dt}$$

This is a second order differential equation.

If $E(t) = E_o \sin \omega t$

The above differential equation becomes

$$L \frac{d^2 I}{dt^2} + R \frac{dI}{dt} + \frac{1}{C} I = \omega E_0 \cos \omega t$$

Its auxiliary equation will be:

$$m^2 + \frac{R}{L} m + \frac{1}{LC} = 0$$

Its roots are as follows:

$$m_1 = -\alpha + \beta$$

$$m_2 = -\alpha - \beta$$

Where

$$\alpha = \frac{R}{2L}$$

$$\beta = \frac{1}{2L} \sqrt{R^2 - \frac{4L}{C}}$$

To obtain the particular solution, assume:

$$I_o = k_1 \cos \omega t + k_2 \sin \omega t$$

Then

$$I \frac{dp}{dt} = -\omega k_1 \sin \omega t + \omega k_2 \cos \omega t$$

And

$$I \frac{d^2 p}{dt^2} = -\omega^2 k_1 \cos \omega t - \omega^2 k_2 \sin \omega t$$

Substituting into equation (b), leads to:

$$\left(-\omega^2 Lk_1 + \omega Rk_2 + \frac{1}{x}k_1\right)\cos\omega t$$

$$+\left(-\omega^2 Lk_2 + \omega Rk_1 + \frac{1}{c}k_2\right)\sin\omega t$$

$$= 2E_o \cos\omega t$$

Hence;

$$\left(\frac{1}{c} - \omega^2 L\right)k_1 + \omega Rk_2 = \omega E_o \qquad\qquad \textbf{(c-1)}$$

And

$$-\omega Rk_1 + \left(\frac{1}{c} - \omega^2 L\right)k_2 = 0 \qquad\qquad \textbf{(c-2)}$$

Let us assume:

$$S = \omega L - \frac{2}{\omega c}$$

Then the above equations are reduced to:

$$-SK_1 + Rk_2 = E_0$$

$$Rk_1 + Sk_2 = 0$$

From these equations, one can obtain:

$$k_1 = \frac{-SE_0}{R^2 + S^2}$$

$$k_2 = \frac{RE_0}{R^2 + S^2}$$

In which, S is called **reactance**, given by the following expression:

$$S = \omega L - \frac{1}{\omega c}$$

Example

Find the current $I(t)$ in a $R - L - c$ circuit with $R = 100$ Ohms, $L = 0.1$ Henery and $c = 10^{-3}$ Farad. The circuit is connected to $E(t) = 155 \sin(377) t$. Assuming zero charge and current when $t = 0$.

Solution

The differential equation representing the given $R - L - c$ circuit is given by:

$$(0.1)\frac{d^2 I}{dt^2} + (100)\frac{dI}{dt} + (1000)I = (155 \times 377)\cos(377)t$$

The reactance

$$S = \omega L - \frac{1}{\omega c}$$

$$= 35$$

And the steady state current

$$I_P - k_1 \cos(377)\, t + k_2 \sin(377)\, t$$

Where:

$$k_1 = \frac{-SE_0}{R^2 + S^2} = \frac{-35 \times 155}{100^2 + 35^2} = -0.483$$

$$k_2 = \frac{RE_0}{R^2 + S^2} = \frac{100 \times 155}{100^2 + 35^2} = 1.381$$

Solve the auxiliary equation:

$$(0.1)m^2 + (100)m + 1000 = 0$$

Leads to the following two roots:

$$m_1 = -10, \quad m_2 = -990$$

Hence the general solution:

$$I(t) = c_1 e^{-(10)\,t} + c_2 e^{-(990)\,t} - 0.483 \ \cos(377)\,t + 1.381 \ \sin(377)\,t$$

Determining c_1 and c_2 from the given initial conditions leads to:

$$I(0) = 0$$

$$Q(0) = 0$$

Differentiating the obtained solution, we get:

$$\frac{dI(t)}{dt} = -(10)c_1 e^{-(10)\,t} - (990)c_2 e^{-(990)\,t}$$
$$+ 0.483 \times 377 \sin(377)\,t + 1.381 \times 377 \cos(377)\,t$$

To apply the condition $Q(0) = 0$, one can use the following expression relating current and charge:

$$I(t) = \frac{1}{L} \left[E(t) - RI(t) - \frac{1}{c} Q(t) \right]$$

Because:

$$I = \frac{dQ}{dt}$$

$$Q = \int I dt$$

We have:

$$E(0) = I(0) = Q(0) = 0$$

Therefore:

$I'(0) = 0$

Hence:

$I'(0) = -10 \ c_1 - 990 \ c_2 + 1.381 \times 377 = 0$

We obtain:

$c_1 = -0.043, \qquad c_2 = 0.526$

Finally:

$$I(t) = -0.043 \ e^{-(10)t} + 0.526 \ e^{-(990)t}$$
$$- 0.483 \cos(377) \ t + 1.381 \sin(377) \ t$$

14. POWER SERIES SOLUTION FOR DIFFERENTIAL EQUATIONS

We will consider a method that is often successful for obtaining solutions for differential equations.

14.1. The Power Series Method

In this section, we review of the basic properties of power series before discussing the power series method [89, 90].

A Power Series

A power series in $(x - a)$ is an infinite series of the form:

$$\sum_{n=0}^{n=\infty} c_n (x - a)^n = c_o + c_1(x - a) + c_2(x - a)^2 + \cdots$$

$c_i, i = 0,1,2,..........,\infty$ are constants called the ***coefficients*** of the series, a is a constant called the ***center*** of the series.

14.2. Solving an Initial Value Problem Using Power Series

The fundamental assumption made in solving a linear differential equation of the following form:

$$f(x, y, y', y'', y''', ...) = 0$$

By the power series method the solution of the differential equation can be expressed in the form of a power series, as follows:

$$y = \sum_{n=0}^{n=\infty} c_n x^n = c_o + c_1 x + c_2 x^2 + \cdots$$

Then, differentiate this solution as many times as the number of derivatives that appear in the differential equation. Collecting the terms involving like powers of x, we then obtain an expression of the following form:

$$k_0 + k_1 x + k_2 x^2 + \cdots = \sum_{n=0}^{n=\infty} k_n c x^n = 0$$

Where $k_i, i = 0,1,2,\ldots\ldots,\infty$ are expressions involving the unknown coefficients $c_i, i = 0,1,2,\ldots\ldots,\infty$. Since the k-series must hold for all values of x in the interval of convergence, all k must be zero. We get system of equation, their solutions lead to the unknown values of $c_i, i = 0,1,2,\ldots\ldots,\infty$.

Example

Using the power series method to solve the following initial value problem:

$$y' = y + x^2, \qquad y(0) = 1$$

Solution

Assume a power series solution of the form:

$$y = \sum_{n=0}^{n=\infty} c_n x^n = c_o + c_1 x + c_2 x^2 + \cdots$$

Differentiate the assumed solution,

$$y' = \frac{d}{dx}\left(\sum_{n=0}^{n=\infty} c_n x^n\right) = c_1 + 2c_2 x + 3c_3 x^2 + 4c_4 x^3 + \cdots$$

Substituting by the assumed solution and its derivative into the differential equation leads to:

$$c_1 + 2c_2 x + 3c_3 x^2 + 4c_4 x^3 + \cdots = \left(c_o + c_1 x + c_2 x^2 + \cdots \right) + x^2$$

Collecting like powers of x yields:

$$\left(c_1 - c_o \right) + \left(2c_2 - c_1 \right)x + \left(3c_3 - c_2 - 1 \right)x^2 + \left(4c_4 - c_3 \right)x^3 + \cdots = 0$$

Equating each of the coefficients to zero, we obtain:

$$\left(c_1 - c_o \right) = 0$$
$$\left(2c_2 - c_1 \right) = 0$$
$$\left(3c_3 - c_2 - 1 \right) = 0$$
$$\left(4c_4 - c_3 \right) = 0$$

From which, we find that:

$$c_1 = c_0$$
$$c_2 = \frac{c_0}{2!}$$
$$c_3 = \frac{c_0 + 2}{3!}$$
$$c_4 = \frac{c_0 + 2}{4!} \quad \ldots \ldots$$

Substituting these values into the series solution assumed we get:

$$y = \sum_{n=0}^{n=\infty} c_n x^n = c_o + c_o x + \left(\frac{c_0}{2!} \right)x^2 + \left(\frac{c_0 + 2}{3!} \right)x^3 + \left(\frac{c_0 + 2}{4!} \right)x^4 + \cdots$$

$$y = \sum_{n=0}^{n=\infty} c_n x^n = c_o + c_o x + \left(\frac{c_0}{2!} \right)x^2 + \left(\frac{c_0 + 2}{3!} \right)x^3 + \left(\frac{c_0 + 2}{4!} \right)x^4 + \cdots$$

Apply the initial condition, to get the unknown constant:

$$c_0 = 1$$

$$y = 1 + x + \left(\frac{1}{2!}\right)x^2 + \left(\frac{3}{3!}\right)x^3 + \left(\frac{3}{4!}\right)x^4 + \cdots$$

Example

Using the power series method to solve the following initial value problem:

$$y'' + xy' + y = 0$$

Solution

Assume a power series solution of the form:

$$y = \sum_{n=0}^{n=\infty} c_n x^n$$

$$y' = \sum_{n=0}^{n=\infty} n c_n x^{n-1}$$

$$y'' = \sum_{n=0}^{n=\infty} n(n-1) c_n x^{n-2}$$

Substituting by the assumed solution and its derivatives into the differential equation leads to:

$$\sum_{n=0}^{n=\infty} n(n-1) c_n x^{n-2} + \sum_{n=0}^{n=\infty} n c_n x^n + \sum_{n=0}^{n=\infty} c_n x^n = 0$$

We use the summation notation in this example to develop the skill in manipulating power series that will be required later on. In order to gather all three power series into a single one, we need to re-write each of the sums in the last equation so that the general term contains the same power of the independent variable.

Consider the first series:

$$S_1 = \sum_{n=0}^{n=\infty} n(n-1) c_n x^{n-2}$$

Assume $k = n - 2 \Rightarrow n = k + 2$.

Then

$$S_1 = \sum_{k=0}^{k=\infty} (k+2)(k+1)c_{k+2}x^k$$

Repeating this process for the second and third summation leads to:

$$\sum_{k=0}^{\infty} (k+2)(k+1)c_{k+2}x^k + \sum_{k=1}^{\infty} kc_k x^k + \sum_{k=0}^{\infty} kc_k x^k = 0$$

Gathering like terms in x gives the following equation:

$$\sum_{k=0}^{\infty} \{(k+2)(k+1)c_{k+2} + (k+1)c_k\}x^k = 0$$

Put

$$\{(k+2)(k+1)c_{k+2} + (k+1)c_k\} = 0$$

Therefore

$$c_{k+2} = -\frac{c_k}{k+2}$$

And so,

$$c_2 = -\frac{c_o}{2}$$

$$c_3 = -\frac{c_1}{3}$$

$$c_4 = -\frac{c_0}{8}$$

....

Therefore, the power series solution can be written in the following form:

$$y = c_0\left(1 - \frac{x^2}{2} + \frac{x^4}{2\times 4} - \frac{x^6}{2\times 4\times 6} + ...\right)$$

$$+ c_1\left(x - \frac{x^3}{3} + \frac{x^5}{3\times 5} - \frac{x^7}{3\times 5\times 7} + ...\right)$$

15. ORDINARY AND SINGULAR POINTS

Definition

Singular Point

Suppose that we have a differential equation of the form:

$$y'' + a(x)y' + b(x)y = 0$$

A point $x = 0$ is an ordinary point when both $a(x)$ & $b(x)$ are analytic at $x = 0$.

If $x = 0$ is not analytic point, then it is called singular points.

Example

Using the power series method to solve the following differential equation:

$$(1 - x^2)y'' - 2xy' + 2y = 0$$

Solution

Re-write the differential equation in the following form:

$$y'' - \frac{2x}{(1 - x^2)}y' + \frac{2}{(1 - x^2)}y = 0$$

From this new form, we observe that:

$$a(x) = \frac{-2x}{(1 - x^2)} = -2x(1 + x^2 + x^2 + ...)$$

$$b(x) = \frac{2}{(1-x^2)} = 2\left(1 + x^2 + x^2 + \ldots\ldots\right)$$

These are two Maclaurin series that converge in the interval $|x| < 1$. Hence, the point $x = 0$ is an ordinary point. So the power series solution will yield the general solution.

The complete solution is left to students as an exercise.

Answer

$$y = c_1 x + c_0\left(1 - x^2 - \frac{x^4}{3} - \frac{x^6}{5} - \ldots\right)$$

Example

Solve the following differential equation:

$$y'' + \frac{2}{x}y' + y = 0$$

Solution

In this example $a(x) = \frac{2}{x}$ is not defined at $x = 0$. Therefore, this point is a singular point. To apply the power series method, re-write the differential equation as follows:

$$xy'' + 2y' + xy = 0$$

Applying the assumed series solution and its derivatives into the differential equation yields:

$$x\sum_{n=2}^{\infty} n(n-1)c_n x^{n-2} + 2\sum_{n=0}^{n=\infty} nc_n x^{n-1} + x\sum_{n=0}^{n=\infty} c_n x^n = 0$$

Put

$$k = n-1 \Rightarrow n = k+1$$

Therefore, the series becomes:

$$2c_1 + \sum_{k=1}^{\infty}\left\{(k+2)k + 2(k+1)c_{k+1} + c_{k-1}\right\}x^k =$$

$$= 2c_1 + \sum_{k=1}^{\infty}\left\{(k+1)(k+2)c_{k+1} + c_{k-11}\right\}x^k = 0$$

Yields,

$$c_{k+1} = -\frac{c_{c-1}}{(k+1)(k+2)}$$

From which, the unknown coefficients can be found. Then, the series solution for the differential equation finally, will be:

$$y = c_0\left(1 - \frac{x^2}{3!} + \frac{x^4}{5!} - \frac{x^6}{7!} - \ldots\right)$$

$$= c_0\left(\frac{\sin x}{x}\right)$$

If we look at the obtained solution, we see that it cannot be written as a power series in x. However, it can be written as a power series in x times a power of x.

This suggests that we should try to find solutions of the form:

$$y = x^r\left(c_0 + c_1x + c_2x^2 + c_3x^3 + \ldots\ldots\right)$$

Where r is some real or complex number whenever $x = 0$ is a singular point of the differential equation.

Regular Singular Point

Suppose that we have a differential equation of the form:

$$y'' + a(x)y' + b(x)y = 0$$

A point $x = 0$ is said to be regular singular point if the following two functions $x a(x)$ & $x^2 b(x)$ have convergent Maclurian series in an open interval containing $x = 0$.

16. THE FROBENIUS METHOD: (CASE 1)

Before studying the Frobenius's method, let us consider the following differential equation:

$$y'' + a(x)y' + b(x)y = 0$$

has regular singular point at $x = 0$, and has a solution of the form:

$$y = x^r \left(c_0 + c_1 x + c_2 x^2 + c_3 x^3 + \ldots \right) = \sum_{n=0}^{\infty} c_n x^{r+n}, x > 0$$

Then

$$y = \sum_{n=0}^{\infty} c_n x^{r+n}$$

$$y' = \sum_{n=0}^{\infty} c_n (r+n) x^{r+n-1}$$

$$y'' = \sum_{n=0}^{\infty} c_n (r+n)(r+n-1) x^{r+n-2}$$

Therefore, the series form for the differential equation takes the following form:

$$\sum_{n=0}^{\infty} c_n (r+n)(r+n-1) x^{r+n-2} + a(x) \sum_{n=0}^{n=\infty} c_n (r+n) x^{n-1} + b(x) \sum_{n=0}^{n=\infty} c_n x^{r+n} = 0$$

Factoring an x and x^2 in the second and third series, respectively, we obtain:

$$\sum_{n=0}^{\infty} c_n \left[(r+n)(r+n-1) + (r+n)x a(x) + x^2 b(x) \right] x^{r+n-2} = 0$$

As $x = 0$ is a regular singular point, both $a(x)x$ and $b(x)x^2$ can be expressed as convergent power series in x:

$$xa(x) = a_o + a_1x + a_2x^2 + a_3x^3 + \cdots\cdots\cdots$$

$$x^2b(x) = b_o + b_1x + b_2x^2 + b_3x^3 + \cdots\cdots\cdots$$

But $n \geq 0$, so x^{r-2} is the smallest power of x. Since, the coefficients of a power series whose sum is zero must vanish, we have, for $n = 0$:

$$c_o\left[r(r-1) + a_o r + b_o\right] = 0$$

By hypothesis $c_o = 1$, we obtain the **indicial equation** has two roots. These roots are called the **exponents** of the differential equation.

Three possible cases arise, they are:

•**Case (1):** r_1 and r_2 differ but not by an integer

•**Case (2):** $r_1 = r_2$

•**Case (3):** r_1 and r_2 differ by a non zero integer.

In what follows, we will study the three different cases, in some details.

Case (1): r_1 and r_2 differ but not by an integer

Let the roots of the identical equation differ, but not by an integer. Then, Frobenius's method will yield two solutions, they are:

$$y_1(x) = x^{r_1}\left(c_o + c_1x + c_2x^2 + c_3x^3 + \cdots\cdots\right), \qquad c_o = 1$$
$$y_2(x) = x^{r_2}\left(c_o^* + c_1^*x + c_2^*x^2 + c_3^*x^3 + \cdots\cdots\right) \qquad c_o^* = 1$$

The two solutions are linearly independent.

Example

Solve the following differential equation:

$$xy'' + \frac{1}{2}y' - y = 0 \qquad x > 0$$

Solution

Assume a series solution of the form:

$$y = x^r \left(c_0 + c_1 x + c_2 x^2 + c_3 x^3 + \ldots \ldots \right) = \sum_{n=0}^{\infty} c_n x^{r+n}, x > 0$$

Then

$$y = \sum_{n=0}^{\infty} c_n x^{r+n}$$

$$y' = \sum_{n=0}^{\infty} c_n (r+n) x^{r+n-1}$$

$$y'' = \sum_{n=0}^{\infty} c_n (r+n)(r+n-1) x^{r+n-2}$$

Therefore, the series form for the differential equation takes the following form:

$$\sum_{n=0}^{\infty} c_n (r+n)(r+n-1) x^{r+n-1} + \frac{1}{2} \sum_{n=0}^{n=\infty} c_n (r+n) x^{r+n-1} - \sum_{n=0}^{n=\infty} c_n x^{r+n}$$

$$= \sum_{n=0}^{n=\infty} c_n (r+n)(r+n-1) x^{r+n-1} + \frac{1}{2} \sum_{n=0}^{n=\infty} c_n (r+n) x^{r+n-1}$$

Let $k = n+1$, then $n = k$, yields:

Therefore,

$$-\sum_{n=0}^{\infty} c_{n-1} x^{r+n-1} = c_o \left[r(r-1) + \frac{1}{2} r \right] x^{r-1}$$

$$+ \sum_{n=1}^{\infty} \left\{ (r+n) \left[(r+n-1) + \frac{1}{2} \right] c_n - c_{n-1} \right\} x^{r+n-1} = 0$$

The indicial equation is:

$$r^2 - \frac{1}{2} r = 0$$

Its solution leads to:

$$r^2 - \frac{1}{2}r = 0$$

$$r = 0$$

$$r = \frac{1}{2}$$

For the first root:

$$\sum_{n=1}^{\infty} \left\{ (n)\left[(n-1) + \frac{1}{2} \right] c_n - c_{n-1} \right\} x^{r+n-1} = 0$$

Then

$$c_n = \frac{c_{n-1}}{n(n-0.5)} \qquad n \geq 1$$

The corresponding solution is given as:

$$y_1(x) = c_0 + \frac{2c_0}{1!}x + \frac{2^2 c_0}{2!(1\times 3)}x^2 + \frac{2^3 c_0}{3!(1\times 3\times 5)}x^3 + \cdots$$

$$= c_0 \left(\sum_{n=0}^{\infty} \frac{(2x)^n}{n!(1\times 3\times 5\times \cdots (2n-1))} \right)$$

$$= c_0 \left(\sum_{n=0}^{\infty} \frac{(2)^n (2x)^n}{(2\times 4\times 6\times \cdots 2n)(1\times 3\times 5\times \cdots \times (2n-1))} \right)$$

$$= c_0 \left(\sum_{n=0}^{\infty} \frac{(4x)^n}{(2n!)} \right)$$

$$= c_0 \left(\sum_{n=0}^{\infty} \frac{\left(2\sqrt{x}\right)^{2n}}{(2n!)} \right)$$

$$= c_0 \cosh\left(2\sqrt{x}\right)$$

Following the same procedure for the next root, we get:

$$c_n = \frac{c_{n-1}}{n(n+0.5)}$$

The corresponding solution is given as:

$$y_2(x) = c_0 \sqrt{x} \left(\sum_{n=0}^{\infty} \frac{(2x)^n}{(n!)(1 \times 3 \times 5 \times \cdots \times (2n+1))} \right)$$

$$= c_0 \sqrt{x} \left(\sum_{n=0}^{\infty} \frac{(2\sqrt{x})^{2n+1}}{(2n+1)!} \right)$$

$$= \frac{c_0}{2} \sinh(2\sqrt{x})$$

Case (2): $r_1 = r_2$

Example

Solve the following differential equation:

$$y'' + y' + \frac{1}{4x^2} y = 0 \qquad x > 0$$

Solution

Assume a series solution of the form:

$$y = \sum_{n=0}^{\infty} c_n x^{r+n}, x > 0$$

Then

$$y = \sum_{n=0}^{\infty} c_n x^{r+n}$$

$$y' = \sum_{n=0}^{\infty} c_n (r+n) x^{r+n-1}$$

$$y'' = \sum_{n=0}^{\infty} c_n (r+n)(r+n-1) x^{r+n-2}$$

Therefore, the series form for the differential equation takes the following form:

$$\sum_{n=0}^{\infty} c_n (r+n)(r+n-1)x^{r+n-2} + \sum_{n=0}^{n=\infty} c_n (r+n)x^{r+n-1} + \sum_{n=0}^{n=\infty} \frac{1}{4}c_n x^{r+n-2} = 0$$

Let $k = n+1$. Then,

$$\sum_{k=1}^{\infty} c_{k-1}(r+k-1)x^{r+k+2} = \sum_{n=1}^{\infty} c_{n-1}(r+n-1)x^{r+n-2} = 0$$

The indicial equation is:

$$r(r-1) + \frac{1}{4} = 0$$

Its solution leads to:

$$r_1 = r_2 = \frac{1}{2}$$

Substituting into the series, we get:

$$\sum_{n=1}^{\infty} \left\{ c_n \left[(n+0.5)(n-0.5) + 0.25 \right] + c_{n-1}(n-0.5) \right\} x^{n-\frac{3}{2}}$$

$$\sum_{n=1}^{\infty} \left\{ n^2 c_n + c_{n-1}(n-0.5) \right\} x^{n-\frac{3}{2}} = 0$$

This leads to the following recurrence relation:

$$c_n = -\frac{(n-0.5)c_{n-1}}{n^2} \qquad n \geq 1$$

The corresponding solution will be:

$$y_1(x) = \sqrt{x}\left(c_0 + \frac{c_0}{2}x + \frac{3c_0}{2^2 \times 2^2}x^2 - \frac{3 \times 5 c_0}{3^2 \times 2^2 \times 3^2}x^3 + \cdots \right)$$

$$= \sqrt{x}\left(\sum_{n=0}^{\infty} \frac{(2n)!}{(n!)^3}\left(\frac{-x}{4} \right)^n \right)$$

SUPPLEMENTARY PROBLEMS

Problem (1)

Solve $\dfrac{dy}{dx} = (x + y)^2$ with y(0) = 0.5.

Find y at x = 0.6, 0.8, 1.0 and 1.2, using the following methods:

    (1) Taylor method

    (2) Euler method

    (3) Modified Euler method

    (4) Runge-Kutta method

Then, compare the results.

Problem (2)

Solve $\dfrac{dy}{dx} = \sqrt{(x + y)}$ with y(0.40) = 0.41.

Find y at x = 0.2, 0.4, 0.6 and 0.8, using the following methods:

    (1) Taylor method

    (2) Euler method

    (3) Modified Euler method

    (4) Runge-Kutta method

Then, compare the results.

Problem (3)

Solve $\dfrac{dy}{dx} = \left(3x + y^2\right)$ with y(0) =1.

Find y at $x = 0.2, 0.4, 0.6$ and 0.8, using the following methods:

    (1) Taylor method

    (2) Euler method

    (3) Modified Euler method

    (4) Runge-Kutta method

Then, compare the results.

Problem (4)

Solve $\dfrac{d^2 y}{dx^2} = (0.5)\sqrt{\left(x + \dfrac{dy}{dx}\right)}$ with $\dfrac{dy(0.4)}{dx} = 0.41$.

Find $\dfrac{dy}{dx}$ at $x = 0.6, 0.8$ and 1.0, using the following methods:

    (1) Taylor method

    (2) Euler method

    (3) Modified Euler method

    (4) Runge-Kutta method

Then, compare the results.

Problem (5)

Solve $\dfrac{dy}{dt} = \sqrt{y} + \sqrt{t}$ with $y(t = 1.0) = 4$.

Find y at $t = 1.2, 1.4$ and 1.6, using the following methods:

    (1) Taylor method

    (2) Euler method

    (3) Modified Euler method

    (4) Runge-Kutta method

Then, compare the results.

Problem (6)

Solve the following system of differential equations:

$$\frac{dx}{dt} = x + \sqrt{y} + 3\sqrt{t}$$

And

$$\frac{dy}{dt} = \sqrt{x} + \sqrt{y} - 3t$$

With $x(t = 0) = 2$ *and* $y(t = 0) = 1$.

Find x *and* y at $t = 0.2, 0.5$ and 0.8, using the following methods:

    (1) Taylor method

    (2) Euler method

    (3) Modified Euler method

    (4) Runge-Kutta method

Then, compare the results.

Problem (7)

Using power series method to solve the following initial value problems:

(a) $y' = y + 1$, $y(0) = 1$

(b) $y' = y + x$, $y(0) = 2$

(c) $y' - 2y = x^2,$ $\qquad y(1) = 1$

(d) $y'' + 4y = 2x,$ $\qquad y(0) = 0, y'(0) = 2$

(e) $y'' + y = 1 + x + x^2,$ $\qquad y(0) = 1, y'(0) = -1$

Problem (8)

Find the general series solution of each of the following differential equation:

(a) $(1 + x^2)y'' + 2xy' - 2y = 0$

(b) $xy'' - xy' + y = e^x,$ $\qquad y(0) = 0, y'(0) = 2$

(c) $xy'' - x^2 y' + (x^2 - 2)y = 0,$ $\qquad y(0) = 0, y'(0) = 1$

(d) $y'' - 2xy' + 4y = 0,$ $\qquad y(0) = 1, y'(0) = 0$

(e) $y'' - xy' + y = -x\cos x,$ $\qquad y(0) = 0, y'(0) = 2$

Problem (9)

Find two linearly independent power series about the ordinary point $x = 0$ that are solutions to the given differential equations:

(a) $y'' - xy' + y = 0$

(b) $y'' - xy' + 2y = 0$

(c) $y'' + x^2 y' + 2xy = 0$

(d) $(1 + x^2)y'' + 2xy' - 2y = 0$

(e) $y'' - xy' + y = -x\cos x$

Problem (10)

Find the general solution to the given differential equations by the Frobenius at $x = 0$.

(a) $y'' + \dfrac{1}{2x} y' + \dfrac{1}{4x} y = 0$

(b) $4xy'' + 2y' + y = 0$

(c) $9x^2 y'' + x^2 y' + 2y = 0$

(d) $x^2 y'' + xy' + (1 + x)y = 0$

(e) $x^2 y'' + \left(x + \dfrac{2}{9} \right) y = 0$

Problem (11)

In the following problems, use the given resistance, capacitance and initial charge RC-circuit. Find an expression for the charge at all time:

(a) $R = 1\Omega \quad C = 1f \quad E = 12V \quad Q(0) = 0$ coulomb

(b) $R = 10\Omega \quad C = 0.001f \quad E = 10\cos 60tV \quad Q(0) = 0$ coulomb

(c) $R = 100\Omega \quad C = 0.0001f \quad E = 100V \quad Q(0) = 1$ coulomb

Problem (12)

The capacitor C in a RC-circuit is charged to 10 volts when the switch is closed. Obtain a differential equation for the capacitor voltage and find the voltage for all times, given that:

(a) $R = 1000\Omega \quad C = 10^{-6} f$

(b) $R = 1000\Omega \quad C = 10^{-3} f$

Problem (13)

Find an expression for the current of a series RL circuit where R = 100 ohms, L = 2 henrys, I(0) =0 and the (emf) satisfies:

$$E = \begin{cases} 6 & 0 \le t \le 10 \\ 7 - e^{10-t} & t \ge 10 \end{cases}$$

Problem (14)

A spring attached by its upper end is stretched 6 inches by a 10-pound weight attached at its lower end. The spring-mass system is suspended in water and is subject to a damping force of five times its velocity. Describe the motion of the spring-mass system and find a particular solution if the weight is drawn down an additional 4 inches and released.

Problem (15)

In the following problems, determine the equation of motion of a mass m attached to a coiled spring with spring constant k, initially displaced a distance, x_o from equilibrium, and released with velocity v_o subject to the following four cases:

(a) No damping or external forces.

(b) A damping constant c but no external forces.

(c) An external force $F_o \sin \omega t$ but no damping

(d) An external force $F_o \sin \omega t$ and damping

$m = 20kg, \quad k = 1000N/m \quad x_o = 1m \quad v_o = 0$

$c = 200kg/s \quad F_o = 1N \quad \omega = 10rad/s$

$m = 25kg, \quad k = 40N/m \quad x_o = 0m \quad v_o = 1m/s$

$c = 20kg/s \quad F_o = 1N \quad \omega = 10rad/s$

CHAPTER 6

Numerical Solution for Partial Differential Equations

Abstract: Partial differential equations have gained great interest due to their efficiency in describing the mathematical model of several physical phenomena and engineering applications. This chapter starts by a general definition for partial differential equations followed by some of the famous analytical methods of solutions and a brief description of major numerical methods for solving partial differential equations.

Keywords: Complex Fourier expansion, Definition of partial differential equation, Half-range expansion, Odd and even function, Types of partial differential equation.

1. INTRODUCTION TO PARTIAL DIFFERENTIAL EQUATIONS

Partial differential equations have gained great interest due to their efficiency in describing the mathematical model of several physical phenomena and engineering applications [91].

Partial differential equations are used to formulate, and thus aid the solution of problems involving functions of several variables.

The differential equation in a certain domain satisfying given conditions is referred to as boundary-value problem. If one or more of the boundaries are not known and moving with time, the problem then is referred to as moving boundary problem [92].

Furthermore, if the governing equation is time independent alongwith boundary condition, then the problem is referred to as free boundary problem.

Partial differential equations are found and used in different fields of science and technology, however despite the great spread of partial differential equations, analytical solutions are restricted to very few simple case problems.

The majority of partial differential equations have no analytical solutions, due to the nature of these equations as well as nonlinear behavior encountered. In the next sub-sections, some of the recent numerical methods will be presented.

Said Gamil Ahmed Sayed Ahmed

The present chapter is an introduction to the partial differential equations, regarding their mathematical behavior, order, degree, linearity, non-linearity and dimensions.

To understand the concept of the above all items, let us take a very simple partial differential equation as an example of illustration.

2. CLASSIFICATION OF THE LINEAR SECOND ORDER P.D.E.

Linear partial differential equation of the second order has the following general form [93]:

$$A\frac{\partial^2 u}{\partial x^2} + 2B\frac{\partial^2 u}{\partial x \partial y} + C\frac{\partial^2 u}{\partial y^2} + D\frac{\partial u}{\partial x} + E\frac{\partial u}{\partial y} + Fu = G \tag{1}$$

Equation (1) in some other references can be re-written as follows:

$$A\frac{\partial^2 u}{\partial x^2} + B\frac{\partial^2 u}{\partial x \partial y} + C\frac{\partial^2 u}{\partial y^2} + D\frac{\partial u}{\partial x} + E\frac{\partial u}{\partial y} + Fu = G \tag{2}$$

According to equation (1), one can classify partial differential equations in a general manner into the following three categories:

Case (1)Hyperbolic

$$\text{If } B^2 - 4AC > 0 \tag{3}$$

Case (2)Parabolic

$$\text{If } B^2 - 4AC = 0 \tag{4}$$

Case (3)Elliptic

$$\text{If } B^2 - 4AC < 0 \tag{5}$$

Some examples for the aforementioned categories are as follows:

$$\frac{\partial^2 u}{\partial t^2} = c^2 \frac{\partial^2 u}{\partial x^2} \quad \text{1-D wave equation} \tag{6}$$

$$\frac{\partial^2 u}{\partial t^2} = c^2 \left(\frac{\partial^2 u}{\partial x^2} + \frac{\partial^2 u}{\partial y^2} \right) \text{2-D wave equation} \tag{7}$$

$$\frac{\partial u}{\partial t} = c^2 \frac{\partial^2 u}{\partial x^2} \text{ 1-D heat equation} \tag{8}$$

$$\frac{\partial u}{\partial t} = c^2 \left(\frac{\partial^2 u}{\partial x^2} + \frac{\partial^2 u}{\partial y^2} \right) \text{2-D heat equation} \tag{9}$$

$$\frac{\partial^2 u}{\partial x^2} + \frac{\partial^2 u}{\partial y^2} = 0 \text{ 2-D Laplace's equation} \tag{10}$$

$$\frac{\partial^2 u}{\partial x^2} + \frac{\partial^2 u}{\partial y^2} = f(x, y) \text{2-D Poisson's equation} \tag{11}$$

Example

Obtain all cases for the coefficients in the following partial differential equation $au_{xx} + bu_{yy} + cu_x + du_y = 0$ to be an elliptic.

Solution

First of all, let us make the following general classification:

1-The given P.D.E. is homogeneous due to the absence of the function, G.

2-The given P.D.E. is linear because all coefficients are constants.

Now, let us write the general form of the second order linear partial differential equation:

$$A\frac{\partial^2 u}{\partial x^2} + 2B\frac{\partial^2 u}{\partial x \partial y} + C\frac{\partial^2 u}{\partial y^2} + D\frac{\partial u}{\partial x} + E\frac{\partial u}{\partial y} + Fu = G \tag{a}$$

And the given equation is:

$$au_{xx} + bu_{yy} + cu_x + du_y = 0 \tag{b}$$

By making a comparison between equations (a) and (b), leads to:

$$A = a$$
$$B = 0 \qquad \qquad \textbf{(c)}$$
$$C = b$$

The general condition is:

$$\text{General formula} = B^2 - 4AC \qquad \qquad \textbf{(d)}$$

By making use of equation (c) into (d), one can get:

$$\text{General formula} = B^2 - 4AC = -4ac \qquad \qquad \textbf{(e)}$$

Now then, we have the following different cases:

1-If $a, b > 0$, this will corresponds to $B^2 - 4AC < 0$ and so the equation will be elliptic type.

2-If a, b will have the opposite sign, this will corresponds to $B^2 - 4AC > 0$ and so the equation will be hyperbolic type.

Example

Show that $au_{xx} + bu_{xy} + cu_y = 0$ is hyperbolic for all possible cases of its constants.

Solution

First of all, let us make the following general classification:

1-The given P.D.E. is homogeneous due to the absence of the function, G.

2-The given P.D.E. is linear because all coefficients are constants.

Now, let us write the general form of the second order linear partial differential equation:

$$A\frac{\partial^2 u}{\partial x^2} + 2B\frac{\partial^2 u}{\partial x \partial y} + C\frac{\partial^2 u}{\partial y^2} + D\frac{\partial u}{\partial x} + E\frac{\partial u}{\partial y} + Fu = G \qquad \qquad \textbf{(a)}$$

And the given equation is:

$$au_{xx} + bu_{xy} + cu_y = 0 \qquad \textbf{(b)}$$

A comparison between equations (a) and (b) leads to:

$$A = a$$
$$B = b \qquad \textbf{(c)}$$
$$C = b$$

The general condition is:

$$\text{General formula} = B^2 - 4AC \qquad \textbf{(d)}$$

By making use of equation (c) and (d), one can get:

$$\text{General formula} = B^2 - 4AC = b^2 > 0 \qquad \textbf{(e)}$$

From equation (e), we see that $B^2 - 4AC$ is always positive for any value positive or negative for b, therefore, the equation is of hyperbolic type.

3. MORE DEFINITIONS

In the following section, we will do some definitions, and to do so let us consider the following partial differential equation:

$$A(x, y, t)\frac{\partial u}{\partial t} + B(x, y, t)\frac{\partial u}{\partial x} + C(x, y, t)\frac{\partial^2 u}{\partial y^2} = f(x, y, t) \qquad \textbf{(12)}$$

In equation (12):

(x, y, t): Spatial and time independent variables, respectively,

A, B, C: Usually known functions,

$u(x, y, t)$: Dependent unknown variable, usually called potential.

3.1. Order Definition

The order is defined as the highest derivative that appears in the given partial differential equation.

3.2. Degree Definition

The degree is defined as the highest power of the highest derivative that appears in the given partial differential equation.

3.3. Dimension Definition

The dimension of the given partial differential equation is defined as the number of the spatial variables it contains.

3.4. Linearity Definition

The given partial differential equation is called linear if the potential and its derivatives are of first order and there are no products involving more than these terms.

4. P.D.E. AND APPLICATIONS

No one can deny the importance of partial differential equations in daily life.

Who can forget ceramic industry?

Who can forget food industry?

Who can forget steel industry?

Who can forget aircraft industry?

One can say that partial differential equations describe many natural laws in our daily life such as conservation of heat and mass.

One can remember that both Poisson and Laplace equations are used to model the steady state temperature distribution in a plane.

The one-dimensional heat equation is used to model time-dependent temperature distribution along heated bar. The wave equation is used to model vibration of beam or slab according to the dimension of the problem.

Last, but not the least, the advection-diffusion equation is used to model transport of a pollutant in a river.

5. BOUNDARY CONDITIONS

There are three different types of boundary conditions that any practical problem will be subjected to, according to the case underhand.

Dirichlet Boundary Condition

$$u(x, y, t) = \overline{u}(x, y, t) \tag{13}$$

Neumann Boundary Condition

$$\frac{\partial u(x, y, t)}{\partial n} = \overline{q}_N(x, y, t) \tag{14}$$

Robin Boundary Condition

$$\frac{\partial u(x, y, t)}{\partial n} = q_R(x, y, t) \tag{15}$$

But no one can say no more than such conditions, may or may not there will be different boundary conditions encountered according to the practical application formulation.

6. WELL-POSED PROBLEM

Any partial differential equation needs boundary and initial condition – if it is time-dependent problem – in the case when a partial differential equation is given associated with its boundary and initial condition, then it is called well-posed problem.

7. Ill-POSED PROBLEM

If it is given partial differential equation with too many conditions, then there will be no solution, and if too few conditions are given, then the solution will be unique. Also the ill-posed problem refers to wrong formulation in boundary or initial conditions.

8. SOLUTION DOMAIN

If someone asks, what does it mean by domain solution?

The answer is so simple, the solution domain is the region in which the solution exists, converges and gets stable. Many authors define the solution domain from different point of views.

In the following, we will define briefly the solution domain from two different point of views, the first one, is from fluid flow point of view and the second is a more general from mathematical point of view.

8.1. Fluid Flow Point of View

8.1.1. Hyperbolic Problems

In the case of hyperbolic problems, the point $P(x_0, t_0)$ can be influenced by the points in the region bounded by the two characteristic lines $x + ct = \text{const}$ and $x - ct = \text{const}$ and for all times $t < t_0$, see Fig. (1).

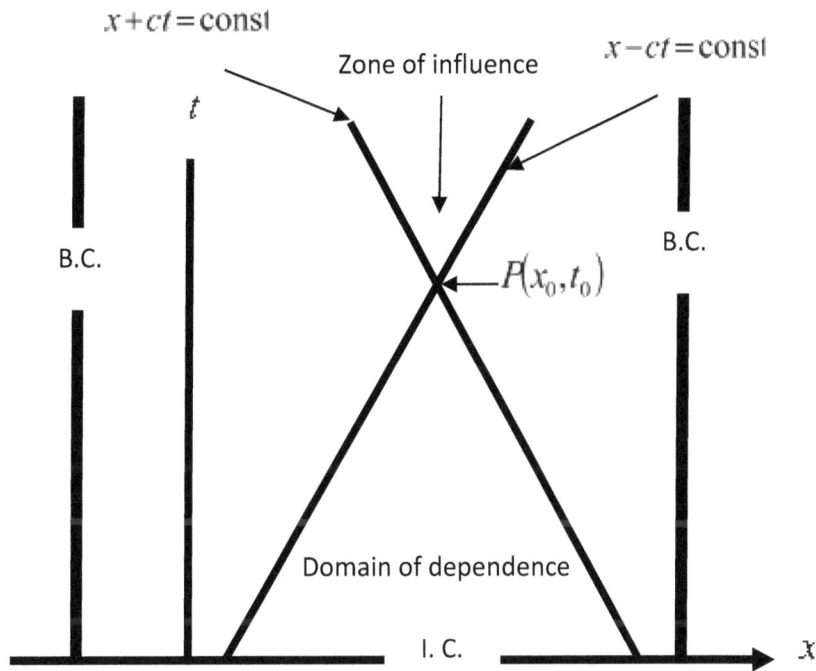

Fig. (1). Domain of Hyperbolic problems.

8.1.2. Parabolic Problems

In the case of parabolic problems, the domain of dependence of the point $P(x_0, t_0)$ is the region $t < t_0$ and the zone of influence is $t > t_0$ see Fig. (2).

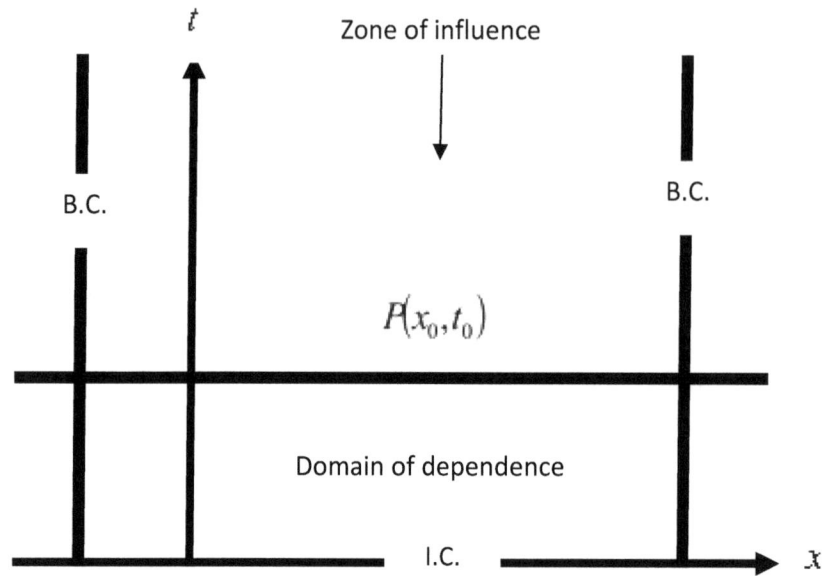

Fig. (2). Domain of parabolic problems.

8.1.3. Elliptic Problems

In the case of elliptic problems, the solution at the point $P(x_0,t_0)$ influences all points within the domain and *vice versa*, see Fig. (**3**).

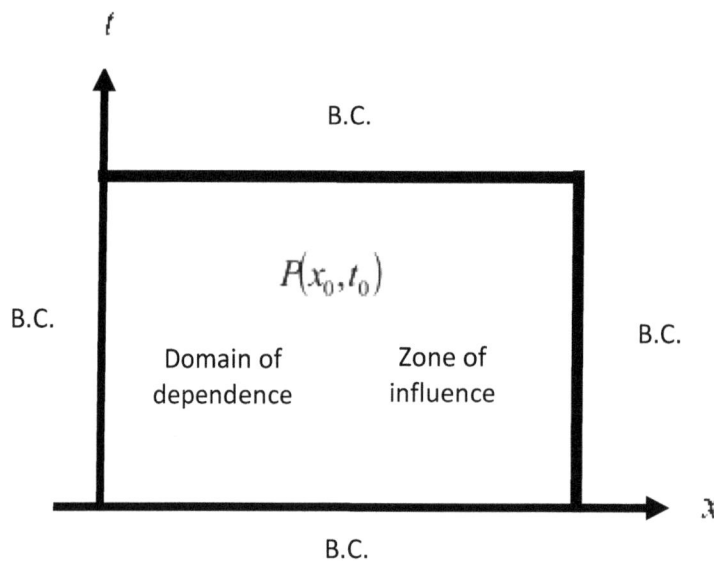

Fig. (3). Domain of Elliptic problems.

9. SERIES FUNDAMENTALS FOR P.D.E.

Partial differential equations have gained the great interest of many researchers due to their efficiency in describing the mathematical model of several physical phenomena and engineering applications.

Partial differential equations are found and used in different fields of science and technology, however despite the great spread of partial differential equations, analytical solutions are restricted to very few simple case problems.

The majority of partial differential equations have no analytical solutions due to the nature of these equations as well as nonlinear behavior encountered.

We will start the numerical treatment of partial differential equation by approximating the derivatives by the finite differences approximations, and then we will follow up the other classical methods of solutions.

10. FORWARD AND BACKWARD FINITE DIFFERENCE FOR 1ST DERIVATIVE

We know that the differential equation in general ordinary or partial consists of dependent and independent variables in addition to derivatives.

If one wants to solve the differential equation, the basic and classical approach is to approximate the derivatives using the classical finite difference approximation and this is what we will do in this section and see what will be the result and how to deal with this result numerically.

Let us write Taylor's expansion for a given function as follows:

$$f(x + \Delta x) = f(x) + (\Delta x)\frac{\partial f}{\partial x} + \frac{(\Delta x)^2}{2!}\frac{\partial^2 f}{\partial x^2} + \frac{(\Delta x)^3}{3!}\frac{\partial^3 f}{\partial x^3} + ... \qquad \textbf{(1)}$$

Equation (1) can be re-written in the following compact form:

$$f(x + \Delta x) = f(x) + \sum_{n=1}^{\infty} \frac{(\Delta x)^n}{n!}\frac{\partial^n f}{\partial x^n} \qquad \textbf{(2)}$$

Re-arrange equation (2) in the following manner:

$$\frac{f(x+\Delta x)-f(x)}{\Delta x}=\frac{1}{\Delta x}\sum_{n=1}^{\infty}\frac{(\Delta x)^{n}}{n!}\frac{\partial^{n} f}{\partial x^{n}}$$

$$\Rightarrow$$

$$\frac{f(x+\Delta x)-f(x)}{\Delta x}=\left[\sum_{n=1}^{\infty}\frac{(\Delta x)^{n-1}}{n!}\frac{\partial^{n} f}{\partial x^{n}}\right] \tag{3}$$

$$=\frac{\partial f}{\partial x}+\frac{(\Delta x)}{2!}\frac{\partial^{2} f}{\partial x^{2}}+\frac{(\Delta x)^{2}}{3!}\frac{\partial^{3} f}{\partial x^{3}}+....$$

By canceling all the terms containing the higher order of (Δx), one can get the following approximation, see Fig. (**4**):

$$\frac{\partial f}{\partial x}=\left(\frac{f(x+\Delta x)-f(x)}{\Delta x}\right)+O(\Delta x) \tag{4}$$

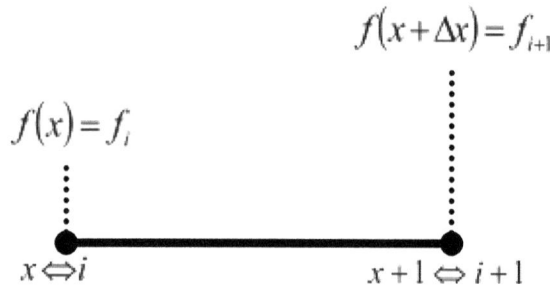

Fig. (4). Function at two successive time steps.

Equation (4) can be re-written in a more general form:

$$\left(\frac{\partial f}{\partial x}\right)_{i}=\left(\frac{f_{i+1}-f_{i}}{\Delta x}\right)+O(\Delta x) \tag{5}$$

Equation (5) is called the first ***forward*** approximation for the first derivative and of order Δx. Assume that one wrote Taylor's expansion as follows:

$$f(x-\Delta x)=f(x)-(\Delta x)\frac{\partial f}{\partial x}+\frac{(\Delta x)^{2}}{2!}\frac{\partial^{2} f}{\partial x^{2}}-\frac{(\Delta x)^{3}}{3!}\frac{\partial^{3} f}{\partial x^{3}}+-.... \tag{6}$$

Again, equation (6) can be re-written in the following compact form:

$$f(x-\Delta x)= f(x)+ \sum_{n=1}^{\infty}\left(\pm\frac{(\Delta x)^{n}}{n!}\frac{\partial^{n} f}{\partial x^{n}}\right) \quad \begin{cases}(+) & for\,even\,(n) \\ (-) & for\,odd\,(n)\end{cases} \qquad (7)$$

Re-arranging equation (7) leads to:

$$\frac{f(x-\Delta x)-f(x)}{\Delta x}= \sum_{n=1}^{\infty}\left(\pm\frac{(\Delta x)^{n-1}}{n!}\frac{\partial^{n} f}{\partial x^{n}}\right) \quad \begin{cases}(+) & for\,even\,(n) \\ (-) & for\,odd\,(n)\end{cases} \qquad (8)$$

Similarly, as done before, one can get another formula for the first approximation for the first derivative as follows:

$$\frac{\partial f}{\partial x}=\left(\frac{f(x)-f(x-\Delta x)}{\Delta x}\right)+O(\Delta x) \qquad (9)$$

Equation (9) in terms of index notations can be re-written as:

$$\left(\frac{\partial f}{\partial x}\right)_{i}=\left(\frac{f_{i}-f_{i-1}}{\Delta x}\right)+O(\Delta x) \qquad (10)$$

Equation (10) is called the first **backward** approximation for the first derivative and of order Δx. Now we derived the first approximation for the derivative of first order, we will turn our attention to the derivation of approximating the second derivatives so as we can be able to solve simple partial differential equations.

11. CENTRAL FINITE DIFFERENCE APPROXIMATION FOR 1ST DERIVATIVE

Let us have two Taylor's expansion of the following forms:

$$f(x+\Delta x)= f(x)+(\Delta x)\frac{\partial f}{\partial x}+\frac{(\Delta x)^{2}}{2!}\frac{\partial^{2} f}{\partial x^{2}}+\frac{(\Delta x)^{3}}{3!}\frac{\partial^{3} f}{\partial x^{3}}+.... \qquad (1)$$

And

$$f(x-\Delta x)= f(x)-(\Delta x)\frac{\partial f}{\partial x}+\frac{(\Delta x)^{2}}{2!}\frac{\partial^{2} f}{\partial x^{2}}-\frac{(\Delta x)^{3}}{3!}\frac{\partial^{3} f}{\partial x^{3}}+-.... \qquad (2)$$

Subtract equation (2) from equation (1), leads to:

$$f(x + \Delta x) - f(x - \Delta x) = 2(\Delta x)\frac{\partial f}{\partial x} + 2\frac{(\Delta x)^3}{3!}\frac{\partial^3 f}{\partial x^3} + \dots \tag{3}$$

Neglecting all terms starting from the term containing $(\Delta x)^3$, gives:

$$\frac{\partial f}{\partial x} = \left(\frac{f(x + \Delta x) - f(x - \Delta x)}{2(\Delta x)}\right) + O(\Delta x)^2 \tag{4}$$

Equation (4) can be written in terms of index notations as follows:

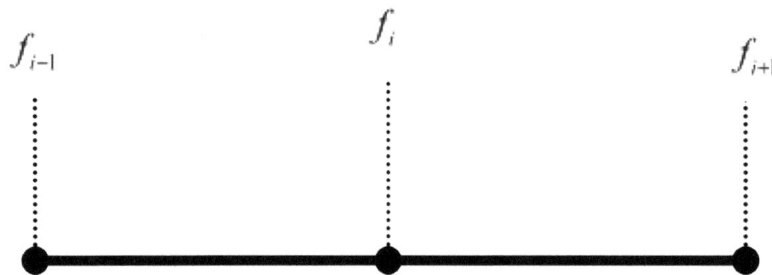

Fig. (5). Function at three points.

$$\left(\frac{\partial f}{\partial x}\right)_i = \left(\frac{f_{i+1} - f_{i-1}}{2(\Delta x)}\right) + O(\Delta x)^2 \tag{5}$$

Equation (5) is called the ***Central difference approximation*** for the first derivative and of order $(\Delta x)^2$.

12. FORWARD AND BACKWARD FINITE DIFFERENCE APPROXIMATION FOR 2$^\text{ND}$ DERIVATIVE

In the last three sections, we derived the approximation of the first derivative using forward, backward and central differences.

In this section, we will derive the approximation of the second derivatives based on Taylor's expansion. Assume that we have two Taylor expansions of the following successive forms:

$$f(x + \Delta x) = f(x) + (\Delta x)\frac{\partial f}{\partial x} + \frac{(\Delta x)^2}{2!}\frac{\partial^2 f}{\partial x^2} + \frac{(\Delta x)^3}{3!}\frac{\partial^3 f}{\partial x^3} + \ldots \qquad (1)$$

And

$$f(x + 2\Delta x) = f(x) + (2\Delta x)\frac{\partial f}{\partial x} + \frac{(2\Delta x)^2}{2!}\frac{\partial^2 f}{\partial x^2} + \frac{(2\Delta x)^3}{3!}\frac{\partial^3 f}{\partial x^3} + \ldots \qquad (2)$$

Multiplying equation (1) by 2 and subtracting the result from equation (2), leads to:

$$-2f(x + \Delta x) + f(x + 2\Delta x) = -f(x) + (\Delta x)^2\frac{\partial^2 f}{\partial x^2} + (\Delta x)^3\frac{\partial^3 f}{\partial x^3} + \ldots \qquad (3)$$

Neglect all terms starting from the term containing $(\Delta x)^3$, this gives:

$$\frac{\partial^2 f}{\partial x^2} = \left(\frac{f(x + 2\Delta x) - 2f(x + \Delta x) + f(x)}{(\Delta x)^2}\right) + O(\Delta x) \qquad (4)$$

Equation (4), can be written in terms of index notations as follows:

$$\left(\frac{\partial^2 f}{\partial x^2}\right)_i = \left(\frac{f_{i+2} - 2f_{i+1} + f_i}{(\Delta x)^2}\right) + O(\Delta x) \qquad (5)$$

Equation (5) is called the ***forward difference approximation*** for the first derivative and of order $(\Delta x)^2$.

$$\left(\frac{\partial^2 f}{\partial x^2}\right)_i = \left(\frac{f_i - 2f_{i-1} + f_{i-2}}{(\Delta x)^2}\right) + O(\Delta x) \qquad (6)$$

Now we are ready to start numerical solution of different types of partial differential equations. The major and basic idea is to replace the differential equation with its associated boundary and initial conditions to a linear system of algebraic equations solved by any numerical method to solve the resultant linear system of algebraic equations.

13. NUMERICAL SOLUTION OF ELLIPTIC P.D.E.

The direct example of elliptic partial differential equation is Laplace equation in two and three dimensions, and the state equation in both two and three dimensions takes the following form:

Case (1): 2-D state equation

$$\nabla^2 u = \frac{\partial^2 u}{\partial x^2} + \frac{\partial^2 u}{\partial y^2} = 0 \tag{1}$$

Case (2): 3-D state equation

$$\nabla^2 u = \frac{\partial^2 u}{\partial x^2} + \frac{\partial^2 u}{\partial y^2} + \frac{\partial^2 u}{\partial z^2} = 0 \tag{2}$$

Before continuing, let us remind the reader to the notation ∇^2 that is called Laplacian operator, and it takes two different forms in Cartesian coordinates, depending on the dimension of the problem under hand.

Case (1): Laplacian in 2-D

$$\nabla^2 = \left(\frac{\partial^2}{\partial x^2} + \frac{\partial^2}{\partial y^2} \right) \tag{3}$$

Case (2): Laplacian in 3-D

$$\nabla^2 = \left(\frac{\partial^2}{\partial x^2} + \frac{\partial^2}{\partial y^2} + \frac{\partial^2}{\partial z^2} \right) \tag{4}$$

13.1. Basic Idea of Solution

Suppose that it is required to solve Laplace equation over a domain bounded by a fixed boundary. Over the boundary, boundary conditions are associated.

The main idea of solution can be summarized as follows:

1-Divide the whole domain by vertical and horizontal lines usually called – Grid – the intersection between any vertical and horizontal line is called node.

2-Replace all the derivatives appearing in the given P.D.E. into the corresponding differential equation.

3-Substituting by differences in the P.D.E., leads to an algebraic equation.

4-By applying the resultant algebraic equation in step 3 to all nodes over the grid we will get a number of algebraic equations that is equal to the number of nodes.

5-Use any numerical method to solve the resultant system of equations.

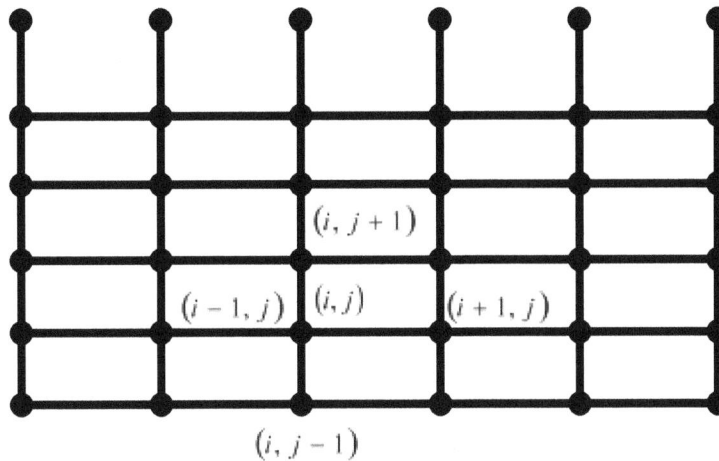

Fig. (6). Grid distribution.

Important Notes

The index *i* represents the node number in the horizontal direction.

The index *J* represents the node number in the vertical direction.

Now let us write the difference approximation for both second derivatives, respectively:

$$\frac{\partial^2 u}{\partial x^2} = \frac{u_{i+1,j} - 2u_{i,j} + u_{i-1,j}}{(\Delta x)^2} \tag{5}$$

$$\frac{\partial^2 u}{\partial y^2} = \frac{u_{i,j+1} - 2u_{i,j} + u_{i,j-1}}{(\Delta y)^2} \tag{6}$$

By making use of equations (5) and (6) into Laplace equation, leads to:

$$\frac{\partial^2 u}{\partial x^2} + \frac{\partial^2 u}{\partial y^2} = \frac{u_{i+1,j} - 2u_{i,j} + u_{i-1,j}}{(\Delta x)^2} + \frac{u_{i,j+1} - 2u_{i,j} + u_{i,j-1}}{(\Delta y)^2} = 0 \qquad (7)$$

Equation (7) is the differential equation corresponding to Laplace equation. This equation in case of unequal space grid distance in both directions takes the following form:

General form $\Delta x \neq \Delta y$

$$\frac{u_{i+1,j} - 2u_{i,j} + u_{i-1,j}}{(\Delta x)^2} + \frac{u_{i,j+1} - 2u_{i,j} + u_{i,j-1}}{(\Delta y)^2} = 0 \qquad (8)$$

Special form $\Delta x = \Delta y$

$$u_{i+1,j} + u_{i-1,j} + u_{i,j+1} + u_{i,j-1} - 4u_{i,j} = 0 \qquad (9)$$

Before continuing and dealing with numerical example, let us take a tour to different types of boundary conditions. Boundary conditions have three different types, they are, Neumann, Dirichlet and Robin.

13.2. First Kind Dirichlet

In this type, the potential is known at all point of the boundary. The word potential refers mainly to the known type, which can be temperature, twisting angle, or something like that.

Example

Find the steady state temperature inside a square plate with pre-scribed boundary condition shown in the attached figure taking $a = b = 1$.

Solution

The domain is a square of side length equals unity, and the grid is constructed in such a way that there is equal space distance in both horizontal and vertical directions, *i.e.*, $\Delta x = \Delta y = 0.25$. The next step of the solution procedure and due to equal grid size, we will make use of the differential equation given by equation (9).

$$u(x, y=b) = 100^0 c$$

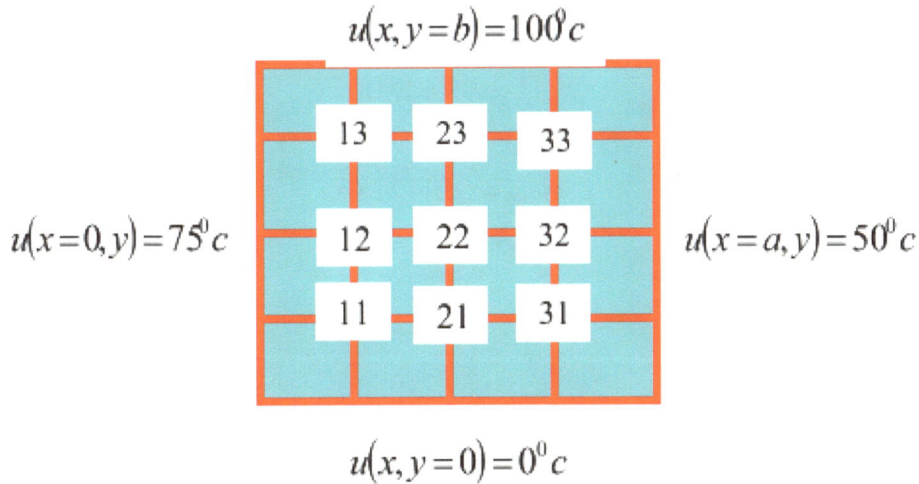

$$u(x=0, y) = 75^0 c \qquad\qquad u(x=a, y) = 50^0 c$$

$$u(x, y=0) = 0^0 c$$

Fig. (7). Domain of interest.

The next step is to number the nodes, in our example, we in a vertical direction as shown in the above Fig. **(7)**. Then apply equation (1) at each node separately, as follows:

At the point $(1,1)$

$$u_{i+1,j} + u_{i-1,j} + u_{i,j+1} + u_{i,j-1} - 4u_{i,j} = 0 \tag{1}$$

At the point $(2,1)$

$$u_{11} + u_{13} + 0 + u_{22} - 4u_{21} = 0 \tag{2}$$

At the point $(3,1)$

$$u_{21} + 50 + 0 + u_{32} - 4u_{31} = 0 \tag{3}$$

At the point $(1,2)$

$$75 + u_{22} + u_{13} + u_{11} - 4u_{12} = 0 \tag{4}$$

By continuing, one can get nine equations in nine unknowns, then using Gauss or Gauss-Jordan or any other method then the final results can be summarized as follows:

$$u_{11} = 43.00061$$
$$u_{12} = 63.21152$$
$$u_{13} = 78.58718$$
$$u_{21} = 33.29755$$
$$u_{22} = 56.11238 \tag{5}$$
$$u_{23} = 76.06402$$
$$u_{31} = 33.88506$$
$$u_{32} = 52.33999$$
$$u_{33} = 69.71050$$

13.3. Second Kind Neumann

In this type, the potential derivative is known at all point of the boundary. The word potential derivative refers mainly to the known whatever its type, may be temperature gradient, twisting angle gradient, or something like that.

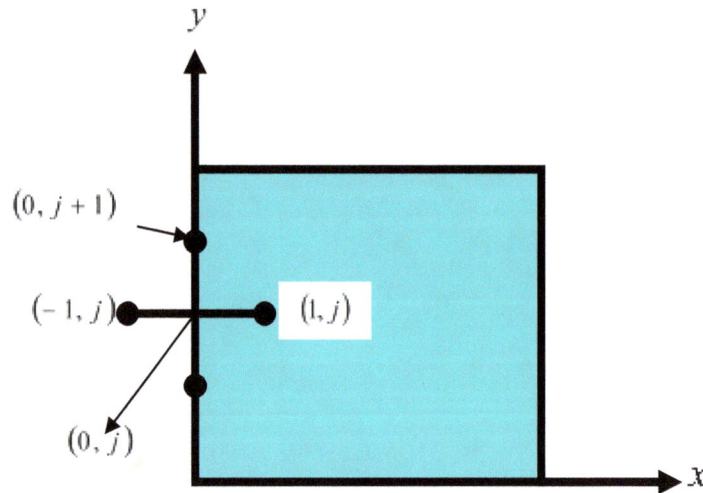

Fig. (8). Dealing with boundary node.

Question will arise if we want to solve the same example before, but with a little bit difference that the left side of the square will be subjected to boundary condition of Neumann's type.

Then the question will be, how can we do in this case?

The answer is that, we take any node on that side and let us denote it as $(0, j)$, and assume that the surrounding nodes will be as shown in Fig. **(8)**, so if we put $i = 0$ in the differential equation given by equation (9), then it will take the following form:

$$u_{1,j} + u_{-1,j} + u_{0,j+1} + u_{0,j-1} - 4u_{0,j} = 0 \qquad (1)$$

$$u_{1,j} + u_{-1,j} + u_{0,j+1} + u_{0,j-1} = 4u_{0,j} \qquad (2)$$

By a deep look to equation (1), one can see an imaginary term and imaginary here does not mean complex but it means it does not exist but we imagine its existence and this term is $\left(u_{-1,j}\right)$.

Now the boundary condition of Neumann's type at the point $(0, j)$ can be written as:

$$\frac{\partial u_{0,j}}{\partial x} = \frac{u_{1,j} - u_{-1,j}}{2(\Delta x)} \qquad (3)$$

The left-hand side of equation (3) is known in advance as a boundary condition therefore, we can find the imaginary value $\left(u_{-1,j}\right)$ as follows:

$$u_{-1,j} = u_{1,j} - 2(\Delta x)\frac{\partial u_{0,j}}{\partial x} \qquad (4)$$

By making use of equation (4) into equation (2), then the later will take the following new form:

$$u_{1,j} + \left[u_{1,j} - 2(\Delta x)\frac{\partial u_{0,j}}{\partial x}\right] + u_{0,j+1} + u_{0,j-1} - 4u_{0,j} = 0 \qquad (5)$$

Simplifying equation (5) will lead to the differential equation in case of Neumann's type boundary condition, taking into consideration that we assumed equal grid size in both directions.

$$2u_{1,j} - 2(\Delta x)\frac{\partial u_{0,j}}{\partial x} + u_{0,j+1} + u_{0,j-1} - 4u_{0,j} = 0 \qquad (6)$$

Example

Find the steady state temperature inside a square plate with pre-scribed boundary condition shown in the attached figure taking $a = b = 1$.

$$u(x, y=b) = 100^0 c$$

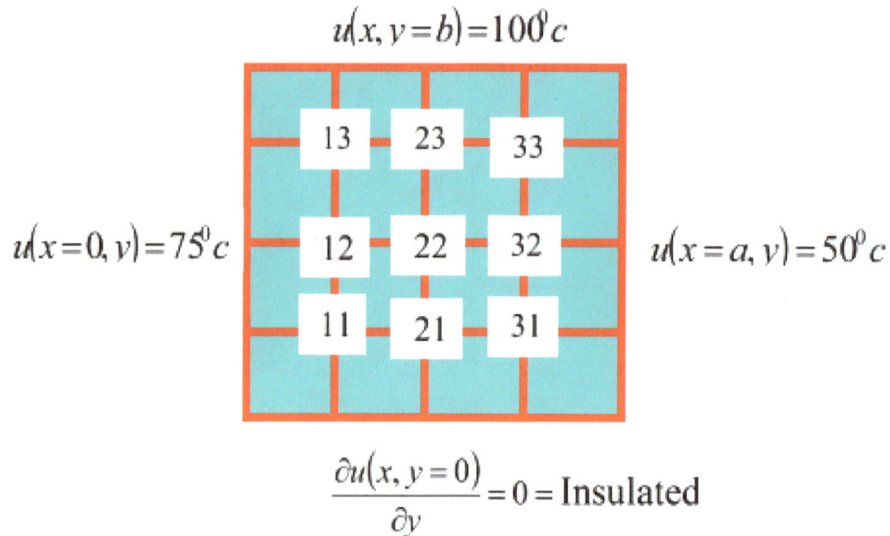

$$u(x=0, y) = 75^0 c$$

13	23	33
12	22	32
11	21	31

$$u(x=a, y) = 50^0 c$$

$$\frac{\partial u(x, y=0)}{\partial y} = 0 = \text{Insulated}$$

Fig. (9). Domain of the problem.

Solution

As we see, the lower side is perfectly insulated, so that, the differential equation will be suitable for solution as follows:

$$u_{i+1,0} + u_{i-1,0} + 2u_{i,1} - 2(\Delta y)\frac{\partial u}{\partial y} - 4u_{i,0} = 0 \qquad \textbf{(1)}$$

Equation (1) is in its general form, but the equation that will be applied at the lower side takes the following form:

$$u_{i+1,0} + u_{i-1,0} + 2u_{i,1} - 4u_{i,0} = 0 \qquad \textbf{(2)}$$

Apply the differential equation (1) at all 6 nodes except the nodes adjacent to the lower side of the square.

For the 3 nodes adjacent to the lower side of the square, apply equation (2). Finally, we get nine equations in nine unknowns. Solving by using Gauss or Gauss-Jordan one can get the following results:

$$u_{10} = 71.91000 \qquad u_{20} = 67.01000 \qquad u_{30} = 59.54000$$
$$u_{11} = 72.81000 \qquad u_{21} = 68.31000 \qquad u_{31} = 60.57000$$
$$u_{12} = 76.01000 \qquad u_{22} = 72.84000 \qquad u_{32} = 64.42000 \qquad \text{(3)}$$
$$u_{13} = 83.41000 \qquad u_{23} = 82.63000 \qquad u_{33} = 74.26000$$

14. IRREGULAR BOUNDARIES

First of all, we do not deal usually with regular boundaries, but in a wide range of applications, the boundaries of the domain is not straight lines but take different curvatures, then a question will arise:

How can we deal with such boundaries?

To have an answer to such a question, let us have a curved boundary, see Fig. (**10**). Suppose that the nearest node to the curved boundary, we see that there are two nodes intersecting with the curved boundary.

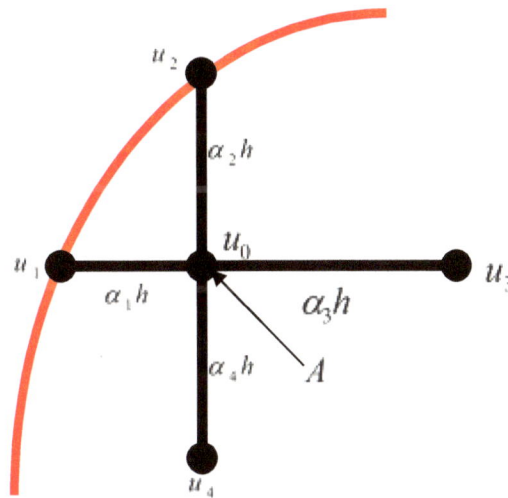

Fig. (10). Node near curved boundary.

Suppose that the node A is called ***unequal-armed star***, let us define the following:

$h = \Delta x = \Delta y$: Arm length and it is assumed equal in both directions

$\alpha_i, i = 1,2,3,4$: Ratio of the arm length

$u_i, i = 0,1,2,3,4$: The potential (temperature) over the grid points

On the straight line joining the points over which the temperatures u_1, u_0 and u_3, we can write the potential derivative using the following formulae:

$$\left(\frac{\partial u}{\partial x}\right)_{1,0} \cong \frac{u_0 - u_1}{\alpha_1 h} \tag{1}$$

$$\left(\frac{\partial u}{\partial x}\right)_{0,3} \cong \frac{u_3 - u_0}{\alpha_3 h} \tag{2}$$

What we want to do?

The answer is so simple, we want to derive a finite difference formula for the point near the curved boundary. Let us start with the second derivative in each direction, separately, as follows:

$$\frac{\partial^2 u}{\partial x^2} = \frac{\partial}{\partial x}\left(\frac{\partial u}{\partial x}\right) \tag{3}$$

Making use of equations (1) and (2) into equation (3), and after some mathematical steps, one can get the following approximation for the second derivative in x-direction, as follows:

$$\frac{\partial^2 u}{\partial x^2} = \frac{\left(\frac{u_3 - u_0}{\alpha_3 h}\right) - \left(\frac{u_0 - u_1}{\alpha_3 h}\right)}{\left(\frac{\alpha_1 + \alpha_3}{2h}\right)} \tag{4}$$

Equation (4) can be re-written in a simplified form:

$$\frac{\partial^2 u}{\partial x^2} = \frac{2}{h^2}\left[\frac{u_1 - u_0}{\alpha_1(\alpha_1 + \alpha_3)} + \frac{u_3 - u_0}{\alpha_3(\alpha_1 + \alpha_3)}\right] \tag{5}$$

By a similar procedure, we can derive the finite differential equation for the second derivative in the transverse y-direction as follows:

$$\frac{\partial^2 u}{\partial y^2} = \frac{2}{h^2} \left[\frac{u_2 - u_0}{\alpha_2(\alpha_2 + \alpha_4)} + \frac{u_4 - u_0}{\alpha_4(\alpha_2 + \alpha_4)} \right] \tag{6}$$

By making use of equations (5) and (6) into Laplace's equation, we get the final finite differential equation corresponding to Laplace's equation over curved boundaries as follows:

$$\frac{2}{h^2} \left\{ \left[\frac{u_1 - u_0}{\alpha_1(\alpha_1 + \alpha_3)} + \frac{u_3 - u_0}{\alpha_3(\alpha_1 + \alpha_3)} \right] + \left[\frac{u_2 - u_0}{\alpha_2(\alpha_2 + \alpha_4)} + \frac{u_4 - u_0}{\alpha_4(\alpha_2 + \alpha_4)} \right] \right\} = 0 \tag{7}$$

$$\left\{ \left[\frac{u_1 - u_0}{\alpha_1(\alpha_1 + \alpha_3)} + \frac{u_3 - u_0}{\alpha_3(\alpha_1 + \alpha_3)} \right] + \left[\frac{u_2 - u_0}{\alpha_2(\alpha_2 + \alpha_4)} + \frac{u_4 - u_0}{\alpha_4(\alpha_2 + \alpha_4)} \right] \right\} = 0 \tag{8}$$

Therefore, equation (8) is the differential equation for Laplace's equation which will be applied at all points near the curved boundaries.

Example

Find the steady state temperature at the indicated grid points on the plate shown in the following figure and its associated boundary conditions.

Solution

First of all, it is important to remember that the whole domain is the upper half of semi-circle, and because of symmetry, we took only the quarter in the first quadrant. Secondly, the points of temperatures u_2, u_3, u_{12} and u_{17} are points of un-equal arms.

Therefore, the ratios $\alpha_i, i = 1,2,3,4$ should be determined. From the analytical geometry, the arm length of u_2 and u_{17} equals $0.899h$ and for the points of u_3 and u_{12} equals $0.5826h$ and the rest of all points are equal and equals to $h = 0.2$. In the following section, we will deal with the aforementioned four points, using equation (7). The final results are listed in Table (**1**).

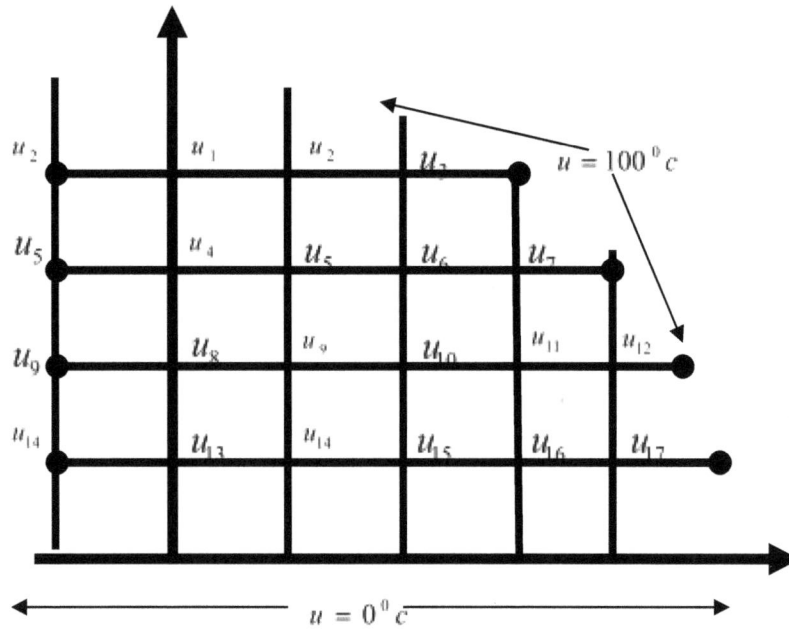

Fig. (11). Grid distribution over curved boundary.

15. NUMERICAL SOLUTION OF PARABOLIC P.D.E.

The second type of partial differential equation we will deal with in a numeric way is the parabolic partial differential equations.

This type of partial differential equations are found in a wide range of engineering and industrial applications. This type of equations are time-dependent and so a care should be taken into consideration specially when dealing with time.

The direct example of this type of partial differential equations is the diffusion problem, which is apparently found in so many disciplines like food industry, ceramic industry, steel industry *etc.*

Let us start with the simplest diffusion case problem in fixed domain, and so the state equation takes the following form:

$$\frac{\partial^2 u}{\partial x^2} = \frac{c\rho}{k}\frac{\partial u}{\partial t} \tag{1}$$

If the grid corresponds to diffusion equation, we will have vertical lines corresponding to different times and the horizontal lines corresponding to the independent variable x. The Index : i will be the variable x and Index : J for the time.

Table 1. Comparison between results.

x	y	Point Number	Finite Difference Results	Analytical Results
0	0.8	1	86.053	85.906
0.2	0.8	2	87.548	87.417
0.4	0.8	3	92.124	92.094
0	0.6	4	69.116	68.807
0.2	0.6	5	70.773	70.482
0.4	0.6	6	75.994	75.772
0.6	0.6	7	85.471	85.405
0	0.4	8	48.864	48.448
0.2	0.4	9	50.436	50.000
0.4	0.4	10	55.606	55.151
0.6	0.4	11	65.891	65.593
0.8	0.4	12	84.189	84.195
0	0.2	13	25.466	25.133
0.2	0.2	14	26.501	26.109
0.4	0.2	15	30.102	29.527
0.6	0.2	16	38.300	37.436
0.8	0.2	17	57.206	57.006

The idea of solution procedure is that, we find the finite difference corresponding to each derivative appearing in the differential equation.

Replace these derivatives by the corresponding differences and finally get the differential equation. The grid that corresponds to the one-dimensional heat diffusion equation is shown in Fig. (**12**).

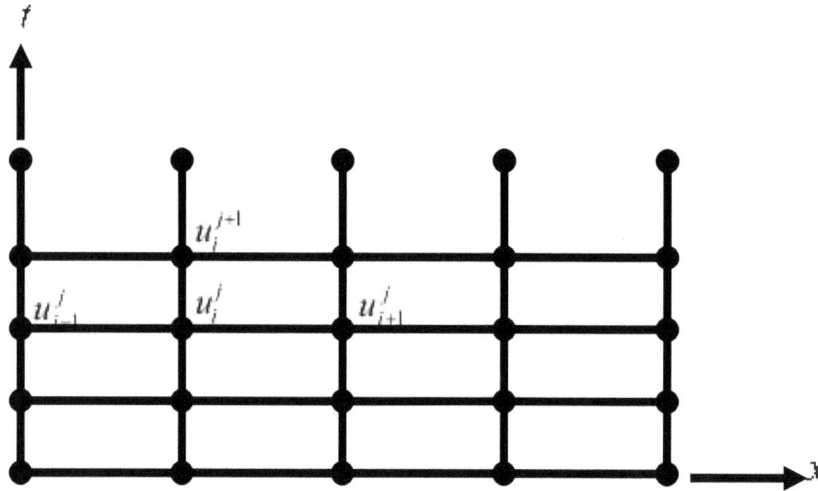

Fig. (12). Grid for 1-D heat diffusion.

The differential equation that corresponds to the second derivative is as follows:

$$\left(\frac{\partial^2 u}{\partial x^2}\right)_{\substack{x=x_i \\ t=t_J}} = \frac{u_{i+1}^J - 2u_i^J + u_{i-1}^J}{(\Delta x)^2} + O(\Delta x)^2 \tag{2}$$

The differential equation that corresponds to the first derivative is as follows:

$$\left(\frac{\partial u}{\partial t}\right)_{\substack{x=x_i \\ t=t_J}} = \frac{u_i^{J+1} - u_i^J}{(\Delta t)} + O(\Delta t) \tag{3}$$

By making use of equations (2) and (3) into equation (1), will lead to the differential equation for one-dimensional heat diffusion equation is as follows:

$$u_i^{J+1} = \left(\frac{k(\Delta t)}{c\rho(\Delta x)^2}\right)[u_{i+1}^J + u_{i-1}^J] + \left[1 - \frac{2k(\Delta t)}{c\rho(\Delta x)^2}\right]u_i^J \tag{4}$$

Notes

u_i^{J+1} Temperature at the point (i) and time $(J+1)$

u_i^{J} Temperature at the point (i) and time (J)

u_{i-1}^{J} Temperature at the point $(i-1)$ and time (J)

u_{i+1}^{J} Temperature at the point $(i+1)$ and time (J)

$\left(\dfrac{k(\Delta t)}{c\rho(\Delta x)^2}\right)$ is a ratio that plays an important role in the results, as we will see later on. This ratio may be one of the following three cases:

Case (I)

$\left(\dfrac{k(\Delta t)}{c\rho(\Delta x)^2}\right) = \dfrac{1}{2}$ In this case, equation (4) will take the following simplified form:

$$u_i^{J+1} = \left(\frac{1}{2}\right)\left[u_{i+1}^{J} + u_{i-1}^{J}\right] \tag{5}$$

This simplification does have any direct effect on the accuracy of the results just it makes the differential equation quite simple.

Case (II)

$\left(\dfrac{k(\Delta t)}{c\rho(\Delta x)^2}\right) < \dfrac{1}{2}$, in this case, we expect an improvement in the obtained results, and there is a very important note here, the size step in x-direction has a direct effect on the accuracy of the solution.

Case (III)

$\left(\dfrac{k(\Delta t)}{c\rho(\Delta x)^2}\right) > \dfrac{1}{2}$, this case only decreases the solution steps but on the other hand we cannot ensure high expected improvement in the obtained results.

Example

A large flat steel plate is 2 cm thickness, if the initial temperature within the plate is given as a function of the distance from one face and is defined as follows:

$$u(x,0) = \begin{cases} 100x & 0 \le x \le 1 \\ 100(2-x) & 1 \le x \le 2 \end{cases}$$

Find the temperature distribution as a function of both space variable and time if both faces are kept at zero temperature.

In this problem, the material properties of the steel are taken as follows:

$k = 0.13\text{cal}/\sec\cdot\text{cm}\cdot°C$

$c = 0.11\text{cal}/g\cdot°C$

$\rho = 0.11g/\text{cm}^3$

Solution

In this case problem, we will solve one-dimensional heat diffusion problem.

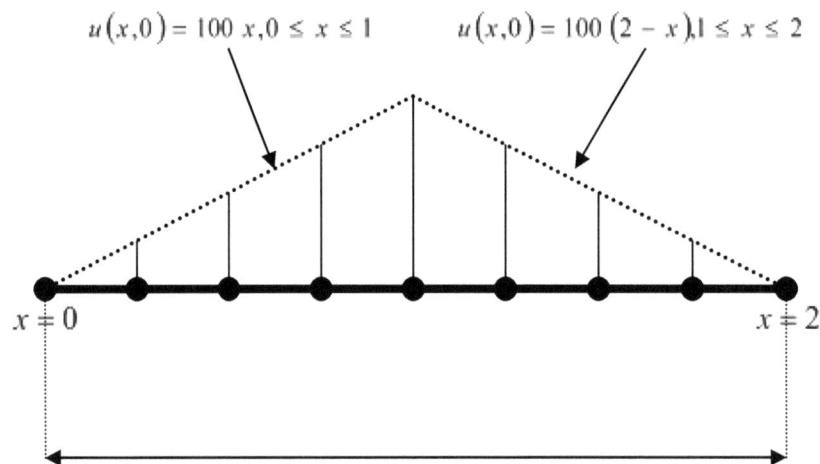

Fig. (13). Problem configuration.

The problem configuration is shown in Fig. (13). The next step is to determine the time step assuming that the step size in x-direction $\Delta x = 0.25$, as follows:

$$\left(\frac{k(\Delta t)}{c\rho(\Delta x)^2} \right) = \frac{1}{2}$$

$$\Rightarrow \qquad\qquad\qquad\qquad\qquad\qquad (1)$$

$$\Delta t = \frac{(0.11)(7.8)(0.25)^2}{(2)(0.13)} = 0.206 \, \text{second}$$

Due to the above assumption regarding time step size, the differential equation will be used as follows:

$$u_i^{J+1} = \left(\frac{1}{2} \right)\left[u_{i+1}^{J} + u_{i-1}^{J} \right] \qquad\qquad (2)$$

Apply the differential equation (2) to all internal nodes shown in Fig. (**13**), and the summary of the computed results will be clearly shown in the following Table (**2**):

Table 2. Computed results.

Time	Calculated Temperature					
	$x = 0.0$	$x = 0.25$	$x = 0.50$	$x = 0.75$	$x = 1.0$	$x = 1.25$
0.000	0	25	50	75	100	75
0.206	0	25	50	75	75	75
0.412	0	25	50	62.5	75	62.5
0.619	0	25	43.75	62.5	62.5	62.5
0.825	0	21.88	43.75	53.12	62.5	53.12
1.031	0	21.88	37.5	53.12	53.12	53.12
1.238	0	18.75	37.5	45.31	53.12	45.31
1.444	0	18.75	32.03	45.31	45.31	45.13
1.650	0	16.16	32.03	38.67	45.31	38.67
1.856	0	16.16	27.34	38.67	38.67	38.67
2.062	0	13.67	27.34	33.01	38.67	33.01

The analytical solution corresponding to the given problem is as follows:

$$u(x,t) = 800 \sum_{n=0}^{\infty} \frac{1}{\pi^2 (2n+1)^2} \cos \frac{(2n+1)\pi(x-1)}{2} e^{-0.3738(2n+1)^2 t} \qquad (3)$$

16. CRANK-NICOLSON METHOD

To understand the Crank-Nicolson method, let us re-write the heat diffusion equation as follows:

$$\frac{\partial^2 u}{\partial x^2} = \frac{c\rho}{k} \frac{\partial u}{\partial t} \qquad (1)$$

Now the differential equation corresponding to time is given as:

$$\frac{\partial u}{\partial t} = \frac{u_i^{j+1} - u_i^{j}}{\Delta t} \qquad (2)$$

And the differential equation corresponding to the second derivative is given as:

$$\frac{\partial^2 u}{\partial x^2} = \frac{1}{2} \left\{ \left(\frac{u_{i+1}^{j} - 2u_i^{j} + u_{i-1}^{j}}{(\Delta x)^2} \right) + \left(\frac{u_{i+1}^{j+1} - 2u_i^{j+1} + u_{i-1}^{j+1}}{(\Delta x)^2} \right) \right\} \qquad (3)$$

By making use of equations (2) and (3) into the heat equation (1), one will get the differential equation and it will take the following form:

$$\frac{1}{2} \left\{ \left(\frac{u_{i+1}^{j} - 2u_i^{j} + u_{i-1}^{j}}{(\Delta x)^2} \right) + \left(\frac{u_{i+1}^{j+1} - 2u_i^{j+1} + u_{i-1}^{j+1}}{(\Delta x)^2} \right) \right\}$$
$$= \left[\frac{c\rho}{k} \right] \left(\frac{u_i^{j+1} - u_i^{j}}{\Delta t} \right) \qquad (4)$$

Notes:

1-The differences that appear in the right-hand side are central differences at the mid time step, *i.e.*, at $t_{j+\frac{1}{2}}$.

2-The first term in the left hand side represents the central difference at time t_{j+1}.

3-The second term in the left hand side represents the central difference at time t_j.

Let us now make the following simplification:

$$r = \frac{k(\Delta t)}{c\rho(\Delta x)^2} \tag{5}$$

By making use of equation (5) into (4), and re-arranging the later, we will get the following equation:

$$-ru_{i-1}^{j+1} + (2+2r)u_i^{j+1} - ru_{i+1}^{j+1} = ru_{i-1}^{j} + (2-2r)u_i^{j} + ru_{i+1}^{j} \tag{6}$$

Equation (6) is called Crank-Nicolson formula for heat diffusion equation.

This method has an important criteria that it is stable for numerical computations for any value of the parameter r taking into consideration that by decreasing the value of the parameter r, better results are obtained with good stability.

Example

Solve

$$\frac{\partial^2 u}{\partial x^2} = \frac{1}{D}\frac{\partial u}{\partial t} \tag{1}$$
$$D = 0.119\text{cm}^2/\sec$$

With the following boundary and initial conditions:

$$u(x,0) = 20 \tag{2}$$

$$u(0,t) = 0 \tag{3}$$

$$u(20,t) = 10 \tag{4}$$

Solution

To solve the example, we first must do some preparations, as follows:

1-Determination of the domain length, and this can be determined from the boundary conditions, it is 20 units.

2-Determination of the step size in x-direction.

3-Determination of the time step.

We do graph the grid of the solution, as shown in Fig. (**14**).

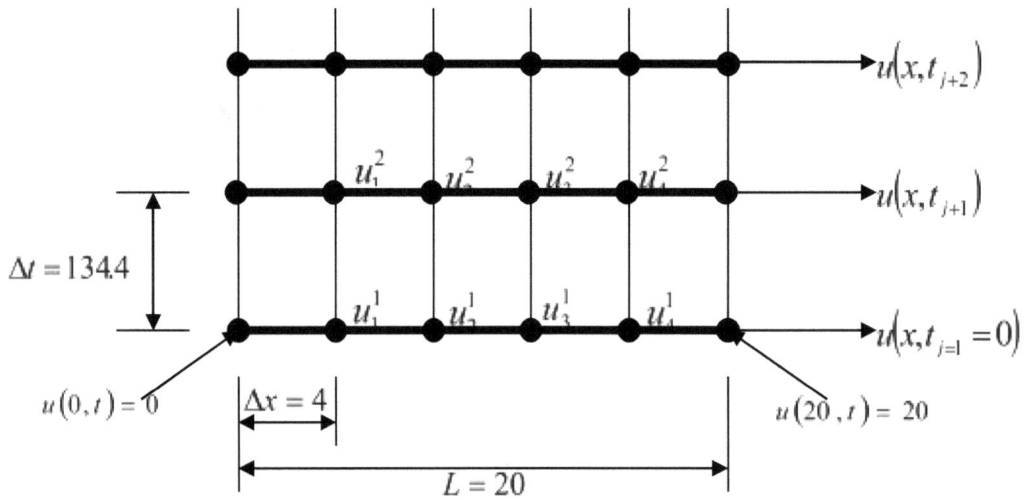

Fig. (14). Problem configuration.

Next let us re-write the relation of the parameter r:

$$r = \frac{k(\Delta t)}{c\rho(\Delta x)^2} \tag{5}$$

Now, and before determination the time step, it is important to choose a suitable value for r, then we will be able to determine the time step.

If one assumes that $r = 1$, then:

$$r = \frac{k(\Delta t)}{c\rho(\Delta x)^2} \tag{6}$$

$$\frac{\partial^2 u}{\partial x^2} = \frac{c\rho}{k} \frac{\partial u}{\partial t}$$

&

$$\frac{\partial^2 u}{\partial x^2} = \frac{1}{D} \frac{\partial u}{\partial t}$$

$$\Rightarrow \tag{7}$$

$$\frac{1}{D} = \frac{c\rho}{k}$$

$$\Rightarrow$$

$$D = \frac{c\rho}{k} = 0.119$$

Substituting by above numeric values, one can get a time step equal to 134.4 sec.

$$-ru_{i-1}^{j+1} + (2+2r)u_i^{j+1} - ru_{i+1}^{j+1} = ru_{i-1}^{j} + (2-2r)u_i^{j} + ru_{i+1}^{j} \tag{8}$$

Put $r = 1$ into equation (8), then it will take the following form:

$$-u_{i-1}^{j+1} + 4u_i^{j+1} - u_{i+1}^{j+1} = u_{i-1}^{j} + ru_{i+1}^{j} \tag{9}$$

Now we are ready to compute the temperature at the next time step taking into consideration that we are now at the time step $t_{j+1} = t_2 = 0 + 134.4 = 134.4$, then apply equation (9) to the points 1,2,3, and 4, respectively, as follows:

$$-0.0 + 4u_1^2 - u_2^2 = 0.0 + 2.0 \tag{10}$$

$$-u_1^2 + 4u_2^2 - u_3^2 = 2.0 + 2.0 \tag{11}$$

$$-u_2^2 + 4u_3^2 - u_4^2 = 2.0 + 2.0 \tag{12}$$

$$-u_3^2 + 4u_4^2 - 10 = 2.0 + 10.0 \tag{13}$$

Repeat the same procedure at each time step, *i.e.*, at each time step four equations like equations (10-14) will be derived, then by any numerical method, the solution will be obtained, see Tables (**3** and **4**).

Table 3. Computed results.

Time in Seconds	Calculated Concentrations					
	$x = 0$	$x = 4$	$x = 8$	$x = 12$	$x = 16$	$x = 20$
0.0	0.000	2.000	2.000	2.000	2.000	10.000
134.4	0.000	0.980	2.019	3.072	5.992	10.000
268.8	0.000	1.070	2.363	4.305	6.555	10.000
403.2	0.000	1.276	2.861	4.762	6.962	10.000
537.6	0.000	1.471	3.065	5.115	7.159	10.000

Table 4. Computed results.

Time in Seconds	Analytical Solution	
	$x = 4$	$x = 12$
134.4	1.078	3.191
268.8	1.108	4.272
403.2	1.340	4.873
537.6	1.543	5.248

17. PARABOLIC P.D.E. IN TWO-DIMENSIONS

The heat diffusion equation is of parabolic type. Let us now start writing the heat equation in two-dimensions, and it will have the following form:

$$\frac{\partial u}{\partial t} = \frac{k}{c\rho}\left(\frac{\partial^2 u}{\partial x^2} + \frac{\partial^2 u}{\partial y^2}\right) \tag{1}$$

The differential equation solution will not differ here than that for one-dimension, except that here an additional term will be added and the grid will be in three-dimensions, two of which represent the xy-plane and the third one represents the time, as shown in Fig. (**15**).

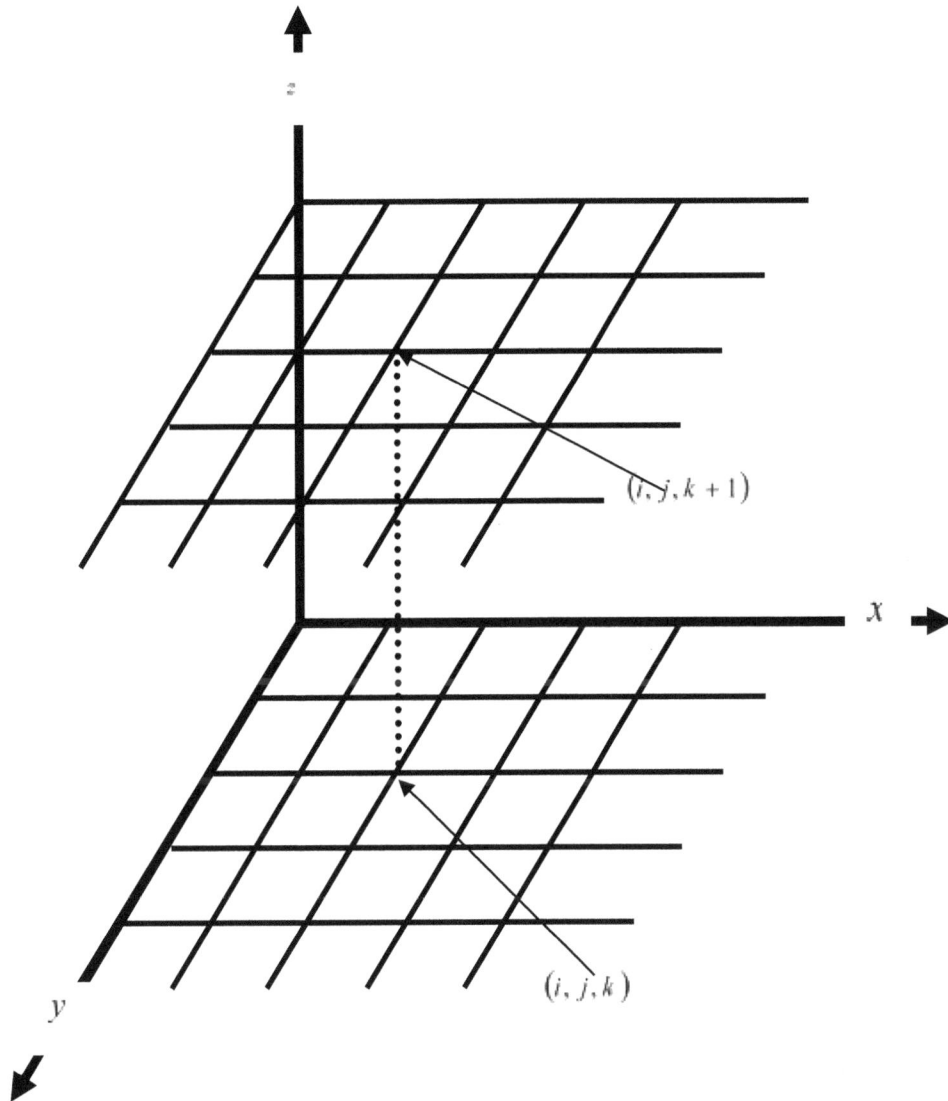

Fig. (15). Two level approximations

Notes

1-The index i will be x-direction.

2-The index j will be y-direction.

3-The index k will be t-direction.

The next step is to represent the derivatives in terms of differences, taking into consideration that we will assume equal space grid in both x and y directions, *i.e.*, $\Delta x = \Delta y$. Another point should be taken into consideration is the parameter r, defined previously as:

$$r = \frac{k(\Delta t)}{c\rho(\Delta x)^2} \tag{2}$$

But for 2-D problems, equation (2) will take another form as follows:

$$r = \frac{k(\Delta t)}{c\rho\{(\Delta x)^2 + (\Delta y)^2\}} \tag{3}$$

Then, the general difference equation for 2-D heat diffusion equation will be of the form:

$$u_{i,j}^{k+1} - u_{i,j}^k = r\left[u_{i+1,j}^k - 2u_{i,j}^k + u_{i-1,j}^k + u_{i,j+1}^k - 2u_{i,j}^k + u_{i,j-1}^k\right] \tag{4}$$

Equation (4) can take the following form:

$$u_{i,j}^{k+1} = r\left[u_{i+1,j}^k + +u_{i-1,j}^k + u_{i,j+1}^k + u_{i,j-1}^k\right] + (1 - 4r)u_{i,j}^k \tag{5}$$

Notes:

1-In case of constant boundary conditions, the maximum allowable value for the parameter r is $r = \frac{1}{4}$.

2-In case of un-equal space grid size $\Delta x \neq \Delta y$, the following relation should be satisfied, that is:

$$\frac{k(\Delta t)}{c\rho\{(\Delta x)^2 + (\Delta y)^2\}} \leq \frac{1}{8} \tag{6}$$

The general Crank-Nicolson formula for two-dimensional heat equation can take the following form:

$$u_{i,j}^{k+1} - u_{i,j}^{k} = \left(\frac{r}{2}\right)\begin{bmatrix}\left(u_{i+1,j}^{k+1} - 2u_{i,j}^{k+1} + u_{i-1,j}^{k+1}\right) + \left(u_{i+1,j}^{k} - 2u_{i,j}^{k} + u_{i-1,j}^{k}\right) \\ \left(u_{i,j+1}^{k+1} - 2u_{i,j}^{k+1} + u_{i,j-1}^{k+1}\right) + \left(u_{i,j+1}^{k} - 2u_{i,j}^{k} + u_{i,j-1}^{k}\right)\end{bmatrix} \quad (7)$$

This means we have to solve system of $M \times N$ linear algebraic equations at each time step, where $M \times N$ represents the number of unknowns in x-direction and $M \times N$ represents the number of unknowns in y-direction.

SUPPLEMENTARY PROBLEMS

Problem (1)

Solve the following partial differential equations:

$(1)\dfrac{\partial^2 z}{\partial x^2}+z=0$

With: At $x=0$ $z=e^y$ and $\dfrac{\partial z}{\partial x}=1$

$(2)\dfrac{\partial^3 z}{\partial x^2 \partial y}+18xy^2+\sin(2x-y)=0$

$(3)\dfrac{\partial^2 z}{\partial x \partial y}=\dfrac{x}{y}+a$

Problem (2)

Which of the following can be solved by the separation of variables method?

$(1)\,a\dfrac{\partial^2 u}{\partial x^2}+b\dfrac{\partial^2 u}{\partial x \partial y}+c\dfrac{\partial^2 u}{\partial y^2}=0$

$(2)\,a\dfrac{\partial^2 u}{\partial x^2}+b\dfrac{\partial^2 u}{\partial y^2}+c\dfrac{\partial u}{\partial x}+d\dfrac{\partial u}{\partial y}=0$

$(3)\,a\dfrac{\partial^2 u}{\partial x \partial y}+bu=0$

Problem (3)

The points of trisection of a string are pulled aside the same distance on opposite sides of the position of equilibrium and the string is released from rest. Derive an expression for the displacement of the string at any time, and show that the mid point of the string always remains at rest.

Problem (4)

Solve completely the equation describing the vibration of a string fixed at both ends. The string released from rest, with initial displacement given by the following equations:

(a) $f(x) = \cos ax$

(b) $f(x) = \sin ax$

(c) $f(x) = \sin ax + \cos ax$

Compare the results at the quarter of the string for the three different cases given above.

Problem (5)

A tightly stretched string fixed at both ends, initially at the position given by

$$f(x) = y_0 \sin^3 \left(\frac{\pi x}{\ell} \right).$$

If the string released from rest, find the displacement at any time and at any position.

Problem (6)

Solve

$$\frac{\partial^2 u}{\partial t^2} = c^2 \frac{\partial^2 u}{\partial x^2}$$

Given that

$$u(0, t) = 0$$

$$u(\ell, t) = 0$$

And

$$u(0 < x < a, t = 0) = \frac{bx}{a}$$

$$u(a < x < \ell, t = 0) = \frac{b(\ell - x)}{(\ell - a)}$$

$$\frac{\partial u(x, t = 0)}{\partial t} = 0$$

Problem (7)

A tightly stretched string fixed at both ends, is distributed from its position of equilibrium by imparting to each of its points and initial velocity of magnitude $f(x)$. Find the solution at any time and any distance.

Problem (8)

An insulated rod of length ℓ has its ends A and B maintained at 0°C and 100°C, respectively till steady state conditions prevail, if the temperature at the end B is suddenly reduced to 0°C and maintained at that temperature. Find the temperature at any distance from the end A.

Problem (9)

An insulated rod of length $\ell = 20$ has its ends A and B maintained at 30°C and 80°C, respectively till steady state conditions prevail. If the temperature at the end A is suddenly raised to 40°C and B reduced to 60°C. Find the temperature at any distance from the end A.

References

[1]   B. J. Martin and D. G. Altman., "Statistics notes: measurement error", *BMJ.,* vol. 313, no. 7059, 1996.

[2]   W. E. Saris and M. Revilla., "Correction for measurement errors in survey research: necessary and possible", Soc. Indicat. Res., vol. 127, pp. 1005-1020, 2015. Doi: 10.1007/s11205-015-1002-x

[3]   W. G. Cochran., Errors of Measurement in Statistics, Technometrics, Taylor & Francis, Ltd. on behalf of American Statistical Association and American Society for Quality. 10: 637-666. Doi: 10.2307/1267450. JSTOR 1267450, November (1968)

[4]   Atkinson and A. Kendall., *An Introduction to Numerical Analysis,* 2nd ed, John Wiley & Sons: New York: p. 20, ISBN 978-0-471-50023-0, 1989.

[5]   J. Stoer, R. Bulirsch, *Introduction to Numerical Analysis,* 3rd ed, Springer-Verlag: Berlin, New York, p. 1, ISBN 978-0-387-95452-3, (2002)

[6]   F. Ayres, E. Mendelson, *Calculus. Schaum's outline series.* 5th ed, McGraw Hill. ISBN 978-0-07-150861-2, 2009.

[7]   R. Wrede and M. R. Spiegel, *Advanced calculus, Schaum's outline series.* 3rd ed. McGraw Hill, ISBN 978-0-07-162366-7, 2010.

[8]   W. F. Hughes and J. A. Brighton, *Fluid Dynamics, Schaum's outline series.* 3rd ed. McGraw Hill, p.160, ISBN 978-0-07-031118-3, 1999.

[9]   R. Penrose, *The Road to Reality*, Vintage books, ISBN 978-00994-40680, 2005.

[10] S. Dineen, *Multivariate Calculus and Geometry, Springer Undergraduate Mathematics Series.* 2nd ed. Springer, ISBN 185-233-472-X, 2001.

[11] N. Bourbaki, *Functions of a Real Variable: Elementary Theory,* Springer. ISBN 354-065-340-6, 2004.

[12] M. A. Moskowitz and F. Paliogiannis., *Functions of Several Real Variables,* World Scientific. ISBN 981-429-927-8, (2011)

[13] W. Fleming., *Functions of Several Variables, Undergraduate Texts in Mathematics.* 2nd ed. Springer, ISBN 0-387-902-066, (1977)

[14] N. I. Achiezer, Theory of approximation, Translated by Charles J. Hyman Frederick Ungar Publishing Co., New York, 1956

[15] A. F. Timan, Theory of approximation of functions of a real variable, ISBN 0-486-67830-X, 1963

[16] C. Hastings, Approximations for Digital Computers, Princeton University Press, 1955.

[17] J. F. Hart, E. W. Cheney, C. L. Lawson, H. J. Maehly, C. K. Mesztenyi, J. R. Rice, H. C. Thacher and C. Witzgall, Computer Approximations, Wiley, Lib. Cong. 67-23326, 1968

[18] L. Fox and I. B. Parker, *Chebyshev Polynomials in Numerical Analysis,* Oxford University Press London, 1968.

[19] S.A. Teukolsky, W. T. Vetterling., *Flannery, BP, Section 5.8. Chebyshev Approximation, Numerical Recipes, The Art of Scientific Computing.* 3rd ed, Cambridge University: New York, Press, ISBN 978-0-521-88068-8, 2007.

[20] W. J. Cody Jr., W. Waite, Software Manual for the Elementary Functions. Prentice-Hall, ISBN 0-13-822064-6, 1980.

[21]   M. Abramowitz, and I. A. Stegun (Eds.), *Handbook of Mathematical Functions with Formulas, Graphs, and Mathematical Tables, 9th printing,* Dover: New York, p. 880, 1972.

[22]   W. H. Beyer., *CRC Standard Mathematical Tables,* 28th ed. CRC Press: Boca Raton, FL, p. 432, 1987.

[23]   R. L. Graham, D. E. Knuth, and O. Patashnik., *Concrete Mathematics: A Foundation for Computer Science,* 2nd ed. Reading, MA: Addison-Wesley, 1994.

[24]   E. T. Whittaker, and G. Robinson., *The Gregory-Newton Formula of Interpolation and An Alternative Form of the Gregory-Newton Formula, §8-9 in The Calculus of Observations, A Treatise on Numerical Mathematics,* 4th ed. Dover; New York, pp. 10-15, (1967)

[25]   R. E. Remez., "Sur le calcul effectif des polynomes d'approximation de Tschebyscheff ", *C. R. Acad. Sci.,* Paris, vol. 199, pp. 337-340, 1934.

[26]   K.-G. Steffens., *The History of Approximation Theory: From Euler to Bernstein,* Birkhauser: Boston ISBN 0-8176-4353-2, (2006)

[27]   T. Erdélyi., "Extensions of the Bloch-Pólya theorem on the number of distinct real zeros of polynomials", J. de théorie des nombres de Bordeaux., vol. 20, pp. 281-287, 2008.

[28]   T. Erdélyi, "The Remez inequality for linear combinations of shifted Gaussians", *Math. Proc. Cambridge Phil. Soc.* Vol. 146, pp. 523-530, 2009.

[29]   L. N. Trefethen., Approximation theory and approximation practice, SIAM, 2013.

[30]   Y. Bugeaud, *Distribution modulo one and Diophantine approximation, Cambridge Tracts in Mathematics.* **193**. Cambridge University Press: Cambridge, 2012. ISBN 978-0-521-11169-0, Zbl 1260.11001.

[31]   J. W. S. Cassels., *An introduction to Diophantine approximation, Cambridge Tracts in Mathematics and Mathematical Physics.* **45**. Cambridge University Press, 1957.

[32]   H. Z. Hassan, A.A. Mohamad, and G. E. Atteia, "An algorithm for the finite difference approximation of derivatives with arbitrary degree and order of accuracy," *J. Comput. Appl. Math.,* vol. 236, no. 10, Pp. 2622-2631, April 2012.

[33]   M. Hazewinkel., Newton-Cotes quadrature formula, Encyclopedia of Mathematics, Springer Science+Business Media B.V. / Kluwer Academic Publishers, 2001. ISBN 978-1-55608-010-4, ed.

[34]   L. F. Richardson., The Approximate Arithmetical Solution by Finite Differences of Physical Problems Involving Differential Equations, with an Application to the Stresses in a Masonry Dam, Philosophical Transactions of the Royal Society A, 210 (459-470), 307-357, doi:10.1098/rsta,1911.0009, JSTOR 90994, (1911)

[35]   W. Romberg., "Vereinfachte numerische Integration, Det Kongelige Norske Videnskabers Selskab Forhandlinger", Trondheim, 28 (7): 30-36, 1955.

[36]   R. Bulirsch, J. Stoer., Handbook Series Numerical Integration, Numerical quadrature by extrapolation, Numerische Mathematik, **9**: 271-278, 1967. doi:10.1007/bf02162420,.

[37]   I.P. Mysovskikh., *Romberg method, in Hazewinkel, Michiel, Encyclopaedia of Mathematics,* Springer-Verlag, 2002. ISBN 1-4020-0609-8.

[38]   S.A Teukolsky, W.T Vetterling and B.P Flannery., *Section 4.3. Romberg Integration, Numerical Recipes: The Art of Scientific Computing,* 3rd ed, Cambridge University Press: New York, 2007 ISBN 978-0-521-88068-8.

[39]   T. J. Bromwich., An Introduction to the Theory of Infinite Series MacMillan & Co. 1908, revised 1926, reprinted 1939, 1942, 1949, 1955, 1959, 1965

[40]   A. Dvoretzky, C. Rogers, and Ambrose., "Absolute and unconditional convergence in normed linear spaces", *Proc. Natl. Acad. Sci. USA.,* vol. 36, no. 3, pp. 192-197, 1950. Doi:10.1073/pnas.36.3.192.

[41]   W. E. Swokowski., *Calculus with analytic geometry,* Alternate ed, Prindle, Weber & Schmidt: Boston, 1983. ISBN 0-87150-341-7, MR0033975.

[42]   S. J. Axler, *Linear Algebra Done Right*, 2nd ed, Springer-Verlag, 1997. ISBN 0-387-98259-0.

[43]   C. L. David, *Linear Algebra and Its Applications,* 3rd ed, Addison Wesley, 2005. ISBN 978-0-321-28713-7.

[44]   C. D. Meyer, *Matrix Analysis and Applied Linear Algebra, Society for Industrial and Applied Mathematics*, SIAM, 2001. ISBN 978-0-89871-454-8.

[45]   D. Poole., *Linear Algebra: A Modern Introduction,* 2nd ed, Brooks/Cole, 2006. ISBN 0-534-99845-3.

[46]   H. Anton., *Elementary Linear Algebra (Applications Version),* 9th ed, Wiley International, 2005.

[47]   S. D. Conte and de B. Carl., *Elementary Numerical Analysis-An Algorithmic Approach*, 3rd Edition, McGraw-Hill, 1980.

[48]   C. E. Froberg., *Introduction to Numerical Analysis,* 2nd Edition, Addison-Wesley, 1981.

[49]   C. F. Gerald, and P.O. Wheatley., *Applied Numerical Analysis*, 5th Edition, Addison-Wesley: Singapore, 1998.

[50]   S. J. Axler., Linear Algebra Done Right (2nd ed.), Springer-Verlag, 1997. ISBN 0-387-98259-0.

[51]   D. C. Lay., *Linear Algebra and Its Applications*, 3rd ed, Addison Wesley, 2005. ISBN 978-0-321-28713-7.

[52]   C. D. Meyer., *Matrix Analysis and Applied Linear Algebra, Society for Industrial and Applied Mathematics,* SIAM, ISBN 978-0-89871-454-8, archived from the original on 2009-10-31, February 15, 2001.

[53]   T. Muir., *A treatise on the theory of determinants, Revised and enlarged by William H. Metzler,* Dover: New York, NY, 1960 [1933]

[54]   D. Poole., *Linear Algebra: A Modern Introduction*, 2nd ed, Brooks/Cole, 2006. ISBN 0-534-99845-3.

[55]   G. B. Price (1947) "Some identities in the theory of determinants", American Mathematical Monthly 54:75-90 MR0019078

[56]   R. A. Horn, C. R. Johnson., *Matrix Analysis,* 2nd ed, Cambridge University Press, 2013. ISBN 978-0-521-54823-6.

[57]   H. Anton., *Elementary Linear Algebra (Applications Version)*, 9th ed, Wiley International, 2005.

[58]   S. J. Leon., *Linear Algebra with Applications,* 7th ed, Pearson Prentice Hall, 2006.

[59]   T.S. Blyth and E.F. Robertson., *Basic Linear Algebra*, Springer, 1998. ISBN 3-540-76122-5.

[60]   J. B. Fraleigh and R. Beauregard, Linear Algebra 2nd edition, p 246, Addison-Wesley ISBN 0-201-11949-8, (1990)

[61]   W. Greub, *Linear Algebra*, 4th edition, pp. 120-5, Springer, 1974. ISBN 0-387-90110-8.

[62] P. C. Shields, *Elementary Linear Algebra* 3rd ed, p 274, Worth Publishers, 1980. ISBN 0-87901-121-1.

[63] G. Strang, *Linear Algebra and Its Applications,* 3rd ed, p 246, Brooks/Cole, 1998. ISBN 0-15-551005-3.

[64] R, A. Calinger., *Contextual History of Mathematics*, Prentice Hall, 1999. ISBN 978-0-02-318285-3.

[65] R.W. Farebrother., Linear Least Squares Computations, STATISTICS: Textbooks and Monographs, Marcel Dekker, ISBN 978-0-8247-7661-9, (1988)

[66] Grcar, Joseph F., How ordinary elimination became Gaussian elimination, Historia Mathematica, 38 (2):163-218, arXiv: 0907.2397, Doi:10.1016/j.hm.2010.06.003, (2011a)

[67] Grcar, Joseph F., Mathematicians of Gaussian elimination (PDF), Notices of the American Mathematical Society, **58** (6): 782-792, (2011b)

[68] Higham, Nicholas, Accuracy and Stability of Numerical Algorithms (2nd ed.), SIAM, ISBN 978-0-89871-521-7, (2002)

[69] V. Gintautas., "Resonant forcing of nonlinear systems of differential equations", *Chaos*, vol. 18, no. (3):033118, 2008. arXiv:0803.2252, Bibcode:2008Chaos,18c3118G, Doi:10.1063/1.2964200. PMID 19045456, (2008)

[70] C. Stephenson., "Topological properties of a self-assembled electrical network *via* ab initio calculation", *Sci. Rep.*, vol. 7, no. 41621, 2017. Doi: 10.1038/srep41621, PMC 5290745, PMID 28155863.

[71] D.K. Campbell., "Nonlinear physics: Fresh breather", *Nature*, vol. 432, no. 7016, pp. 455-456, Nov. 2004. Bibcode: 2004Natur, 432-455C, Doi:10.1038/432455a, ISSN 0028-0836.

[72] D. Lazard., "Thirty years of Polynomial System Solving, and now?", *J. Symbol. Comput.*, vol. 44, no. 3, pp. 222-231, 2009. Doi:10.1016/j.jsc.2008.03.004.

[73] S.A. Billings., *Nonlinear System Identification: NARMAX Methods in the Time, Frequency, and Spatio-Temporal Domains*, Wiley, 2013.

[74] D.K. Faddeev and V.N. Faddeeva., *Computational methods of linear algebra*, Freeman, 1963.

[75] I.S. Berezin and N.P. Zhidkov., *Computing methods*, Pergamon, 1973.

[76] J.M. Ortega., W.C. Rheinboldt., *Iterative solution of non-linear equations in several variables,* Acad. Press, 1970.

[77] A.A. Samarskii and E.S. Nikolaev., *Numerical methods for grid equations,* 1-2 Birkhäuser, 1989.

[78] E. A. Coddington., Norman. Levinson., *Theory of Ordinary Differential Equations.* McGraw-Hill: New York, 1955.

[79] P. Hartman., Ordinary differential equations, Classics in Applied Mathematics, **38**, Philadelphia: Society for Industrial and Applied Mathematics, ISBN 978-0-89871-510-1, MR 1929104, (2002) [1964]

[80] E. L. Ince., *Ordinary Differential Equations,* Dover Publications, New York, ISBN 978-0-486-60349-0, MR 0010757, (1944) [1926]

[81] A. D. Polyanin., V. F. Zaitsev, and A. Moussiaux., *Handbook of First Order Partial Differential Equations,* Taylor & Francis: London, 2002. ISBN 0-415-27267-X, (2002)

[82] D. Zwillinger., *Handbook of Differential Equations*, 3rd ed, Academic Press: Boston, 1997.

[83] T. I. Lakoba,, *Simple Euler method and its modifications (PDF) (Lecture notes for MATH334,* University of Vermont), retrieved 29 February 2012.

[84] U. M. Ascher, L. R. Petzold., *Computer Methods for Ordinary Differential Equations and Differential-Algebraic Equations,* Philadelphia: Society for Industrial and Applied Mathematics, 1998, ISBN 978-0-89871-412-8.

[85] J.C. Butcher., *Numerical Methods for Ordinary Differential Equations,* John Wiley & Sons: New York, 2003, ISBN 978-0-471-96758-3.

[86] E. Hairer, S. P. Nørsett,, Gerhard, Wanner., *Solving ordinary differential equations I: Nonstiff problems,* Springer-Verlag: New York, 1993, ISBN 978-3-540-56670-0.

[87] A. Iserles., *A First Course in the Numerical Analysis of Differential Equations*, Cambridge University Press, 1996, ISBN 978-0-521-55655-2.

[88] J. Stoer., R. Bulirsch., *Introduction to Numerical Analysis*, 3rd ed, Springer-Verlag: Berlin, New York, 2002. ISBN 978-0-387-95452-3.

[89] E. A. Coddington., N. Levinson., *Theory of Ordinary Differential Equations.* McGraw-Hill, New York, 1955.

[90] G. Teschl., *Ordinary Differential Equations and Dynamical Systems,* Providence: American Mathematical Society, 2012, ISBN 978-0-8218-8328-0.

[91] E. C. Zachmanoglou and Dale Thoe W., Introduction to Partial Differential Equations with Applications, Dover Publications, (February 1st 1987)

[92] J. Crank., *Free and Moving Boundary Problems,* Clarendon Press, Oxford, 1984.

[93] Ames. William., *Numerical Methods for Partial Differential Equations,* Academic Press, Inc., 1999.

SUBJECT INDEX

A

Adjoint matrix 89, 90
Algebraic equations 79, 92, 119, 231, 233, 255
  linear 79, 92, 119, 255
  resultant 233
Analytical methods, traditional 150
Analytical solutions 11, 128, 155, 218, 227, 248, 252
Approximations numerical analysis 13, 15, 17, 19, 21, 23, 25, 27, 29, 31, 33, 35, 37, 39, 41, 43, 45, 47, 49, 51, 53, 55, 57, 59, 61, 63, 65, 67, 69

B

Boundaries 151, 218, 224, 232, 234, 236, 239, 240, 241, 242, 249
  curved 239, 240, 241, 242
Boundary and initial condition 224, 249
Boundary conditions 151, 157, 172, 223, 224, 232, 234, 237, 250

C

Coefficient matrix 93, 96, 99, 108, 109, 110, 112, 115, 117, 122
Coefficients 16, 17, 107, 198, 204
  determinant of 107
  unknown 16, 17, 198, 204
Coefficients matrix 94, 124
Column vector 99
Complex fourier expansion 218
Convergence and stability analysis 128
Corresponding elements 84, 87, 88, 118, 119
Corresponding polynomial 16
Crank-Nicolson formula for heat diffusion equation 249
Crank-Nicolson formula for two-dimensional heat equation 255
Crank-Nicolson method 248

D

Damping force 182, 186
Data, equal space 30

Data function 42, 44, 47, 50, 54
  numerical 42, 44
Decimal places 2, 22, 25, 68, 130, 134
Definition of partial differential equation 218
Derivative approximations 11
Derivatives 8, 11, 12, 13, 16, 17, 33, 34, 49, 151, 152, 153, 154, 155, 158, 198, 200, 203, 223, 227, 229, 230, 233, 244, 254
  higher order 12, 13
  partial 8
  second 12, 229, 230, 233
Diagonal elements 86, 112, 115, 119
Differences 1, 2, 3, 4, 22, 23, 25, 26, 27, 30, 34, 151, 164, 171, 230, 233, 236, 248, 249, 254
  absolute 2
  central 230, 248, 249
  first 23, 26
  second 23, 26
Difference tables 11, 22, 27
Differential equation 150, 151, 152, 154, 155, 157, 159, 160, 161, 162, 163, 164, 169, 170, 177, 179, 192, 193, 195, 197, 198, 199, 200, 202, 203, 204, 205, 206, 207, 209, 210, 218, 219, 220, 221, 222, 223, 224, 227, 229, 231, 233, 234, 237, 238, 240, 241, 243, 244, 245, 247, 248
  corresponding 233
  final finite 241
  finite 240
  given 163, 164, 169
  given ordinary 151, 152, 162
  given partial 222, 223, 224
  higher ordinary 150, 151
  linear 179, 197
  linear partial 219, 220, 221
  simple partial 219, 229
  systems of ordinary 150
Differential equation solution 253
Differentiation and integration 11
Diffusion equation 242, 244
  one-dimensional heat 244
Domain of dependence 225, 226

www.ingramcontent.com/pod-product-compliance
Lightning Source LLC
Chambersburg PA
CBHW050817220326
41598CB00006B/242